**This book is to be returned on or before the
last date stamped below or you will be
charged a fine**

Student Support Materials for
Edexcel A Level Maths
Mechanics 1

Author: Ted Graham

William Collins' dream of knowledge for all began with the publication of his first book in 1819. A self-educated mill worker, he not only enriched millions of lives, but also founded a flourishing publishing house. Today, staying true to this spirit, Collins books are packed with inspiration, innovation and practical expertise. They place you at the centre of a world of possibility and give you exactly what you need to explore it.

Collins. Freedom to teach.

Published by Collins
An imprint of HarperCollins *Publishers*
77–85 Fulham Palace Road
Hammersmith
London
W6 8JB

© HarperCollins*Publishers* Limited 2012

10 9 8 7 6 5 4 3 2 1

ISBN – 13: 978-0-00-747606-0

British Library Cataloguing in Publication Data. A Catalogue record for this publication is available from the British Library.

Commissioned by Lindsey Charles and Emma Braithwaite
Project managed by Lindsey Charles
Edited by Susan Gardner
Reviewed by Stewart Townend
Design and typesetting by Jouve India Private Limited
Illustrations by Ann Paganuzzi
Index compiled by Michael Forder
Cover design by Angela English
Production by Simon Moore

Printed and bound in Spain by Graficas Estella

Browse the complete Collins catalogue at:
www.collinseducation.com

This material has been endorsed by Edexcel and offers high quality support for the delivery of Edexcel qualifications.

Edexcel endorsement does not mean that this material is essential to achieve any Edexcel qualification, nor does it mean that this is the only suitable material available to support any Edexcel qualification. No endorsed material will be used verbatim in setting any Edexcel examination and any resource lists produced by Edexcel shall include this and other appropriate texts. While this material has been through an Edexcel quality assurance process, all responsibility for the content remains with the publisher.

Copies of official specifications for all Edexcel qualifications may be found on the Edexcel website - www.edexcel.com

Acknowledgements
The publishers wish to thank the following for permission to reproduce photographs. Every effort has been made to trace copyright holders and to obtain their permission for the use of copyright material. The publishers will gladly receive any information enabling them to rectify any error or omission at the first opportunity.

Cover image: Finance building seen from inner square © Ola Dusegård | istockphoto.com

MIX
Paper from
responsible sources
FSC™ C007454

FSC™ is a non-profit international organisation established to promote the responsible management of the world's forests. Products carrying the FSC label are independently certified to assure consumers that they come from forests that are managed to meet the social, economic and ecological needs of present and future generations, and other controlled sources.

Find out more about HarperCollins and the environment at
www.harpercollins.co.uk/green

Welcome to Collins Student Support Materials for Edexcel A level Mathematics. This page introduces you to the key features of the book which will help you to succeed in your examinations and to enjoy your maths course.

The chapters are organised by the main sections within the specification for easy reference. Each one gives a succinct explanation of the key ideas you need to know.

Examples and answers

After ideas have been explained the worked examples in the green boxes demonstrate how to use them to solve mathematical problems.

Method notes

These appear alongside some of the examples to give more detailed help and advice about working out the answers.

Essential ideas

These are other ideas which you will find useful or need to recall from previous study.

Exam tips

These tell you what you will be expected to do, or not to do, in the examination.

Stop and think

The stop and think sections present problems and questions to help you reflect on what you have just been reading. They are not straightforward practice questions - you have to think carefully to answer them!

Practice examination section

At the end of the book you will find a section of practice examination questions which help you prepare for the ones in the examination itself. Answers with full workings out are provided to all the examination questions so that you can see exactly where you are getting things wrong or right!

Notation and formulae

The notation and formulae used in this examination module are listed at the end of the book just before the index for easy reference.

Contents

Contents

When considering the motion of real objects, there are many complex factors that need to be taken into account. In a first module in mechanics such as this, it is not possible to take account of all of them. To tackle problems, we need to create a mathematical model which provides a simplified view of the situation. For example, the forces acting on a moving car are, in reality, are quite complex. Some of these are:

- The car has four points of contact with the ground and there will be at least one force acting on each of the four tyres.

- There is an air resistance force acting on the car, which will increase with the speed of the car. This force is difficult to deal with mathematically.

- The car has a forward force which is produced because of the rotation of the tyres, caused by the engine, and their interaction with the road. This is also hard to deal with mathematically.

A simple approach to the motion of a car is to model it as a particle travelling on a horizontal surface which has a single tractive or driving force in the direction of motion and a constant resistance force directed backwards. It also has two vertical forces acting on it which balance each other. This allows simple problems to be tackled. As your knowledge of mechanics increases, you will be able to use more sophisticated assumptions which take some of the more complex factors into account.

The mathematical modelling cycle

Mathematical modelling is often a cyclic process, as shown in Figure 1.1. Its first iteration can be a very simple model that enables the first steps to be taken. In a second cycle more factors can be taken into account, so that a refined model is produced in the light of the first solution.

Fig. 1.1
The mathematical modelling cycle.

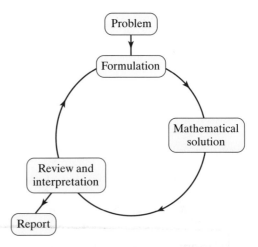

Every modelling problem starts with a problem statement and finishes with a report or solution that is given in terms of the original problem.

The cycle contains three main stages:

- **Formulation stage** In this stage the problem is simplified by making assumptions so that it can be solved using the available mathematical ideas. At the end of this stage the original question will have been converted to a problem that can be solved using mathematical techniques.

- **Mathematical solution** In this stage mathematical techniques are used to solve the problem and to obtain a mathematical answer to the problem.

- **Review and interpretation** In this stage the answers are reviewed to see if they are reasonable. If, for example, one of the assumptions made in the formulation stage has produced an answer that does not make sense in the original context of the problem, it may be necessary to repeat the modelling cycle with a refined set of assumptions.

  Sometimes it may be that, because of the assumptions made, the value obtained is a maximum or minimum value. It is also important to interpret the results obtained in the light of the original problem so that meaningful results can be given to whoever posed the problem with an explanation of any limitations imposed by the assumptions used.

Making assumptions

In this module the main emphasis on mathematical modelling relates to making assumptions.

The bike problem

This is a simple experiment that requires a bike, a piece of string and two people. Tie one end of the string to one of the pedals. One person holds the bike upright, so that it does not fall over, but must allow the bike to move forwards or backwards when the string is pulled. The other person then arranges the pedals so that the one with the string attached is in its lowest possible position. They then stand behind the bike with the string horizontal and then pull it, as shown in Figure 1.2.

Seeing the bike move backwards may be surprising, but it can be explained by making an assumption about the bike. Imagine that the bike is taken to a scrap yard and squashed in a car crusher. The resulting lump of metal and rubber is brought back and placed on a table. If a string is attached to

Essential notes

Many problems can be solved by considering an object to be a particle, but be aware that sometimes this will not give a good solution.

Fig. 1.2
The bike problem.

the lump of metal and pulled, then it will move in the direction of the pull. Modelling the bike as a particle has the same effect, as a particle will move in the direction that it is pulled.

Clearly there is a much more sophisticated solution to this problem, which takes account of the wheels, chain and other parts of the bike, but it requires a higher level of mechanics than is covered in this module.

Modelling objects as particles is simplistic, but it allows solutions to be obtained to problems and as your knowledge of mechanics grows, you can apply more sophisticated approaches.

Standard terms and assumptions

There are a number of modelling terms which describe assumptions used in solving problems in mechanics. The list below contains the terms you are expected to know for the exam.

- **Particle** When an object is modelled as a particle it is considered as if all of its mass is at one single point. This means that its size can ignored and any forces can be treated as if they all act at the same point. Also, any factors due to the rotation of the object can be ignored. In the early sections of this module, almost everything that you meet will be modelled as a particle.

- **Rod** A rod takes account of the size of an object but only in one dimension. A rod would have a length, but no width. When dealing with a rod, the forces act at different points and this factor must be taken into account by taking moments of the forces. Rods are usually **uniform**, which means that the mass of the rod is equally distributed along its length. For a **uniform rod** the weight force can be treated as if it acts at the centre of the rod. It is useful to model objects like ladders as rods. For a **non-uniform rod** the mass is not evenly distributed and the weight cannot be assumed to act at the centre of the rod. For non-uniform rods, the weight is assumed to act at a single point known as the **centre of mass**.

- **Lamina** A lamina has two dimensions, but no thickness. For example, a sheet of metal would normally be modelled as a lamina. A **uniform lamina** would have its mass distributed evenly throughout the lamina. For a uniform rectangular lamina, the weight can be considered as if it acts at the centre of the rectangle. This principle can be applied to laminas with two or more lines of symmetry.

- **Smooth** The term smooth means that there is no friction present. Surfaces can be described as smooth.

- **Rough** The term rough means that friction is present. It does not give any indication of the size of the friction force, but simply that friction is present.

- **Light** This term means that something can be treated as if it has no mass. This term is often used to describe strings or rods that are connected to other bodies that are not light.

- **Inextensible** This term is also often applied to strings. It simply means that the string or other object does not change length and does not stretch in any way. Strings or ropes are often described as being **light and inextensible**.

- **Smooth light pulley** This is often used to describe a pulley in a connected particles problem. It simply means that there is no need to consider what happens to the pulley while solving the problem. In reality a force is needed to make a pulley rotate, but this is usually small compared to the other factors and can be ignored.

- **Peg** In connected particle problems a string may pass over a peg rather than a pulley. This is just like fixed bar that does not rotate or move. With a **smooth peg** there is no friction between the peg and the string.

- **No air resistance** In many of the situations that you consider involving moving objects, you will ignore air resistance.

- **Rigid body** This term refers to any object that has a physical size, unlike a particle, and does not change shape. A rod is probably the simplest example of a rigid body. A lamina is also a rigid body.

- **Bead** A bead is a particle with a hole in it, so that it can be threaded onto a **wire**. The movement of the bead can then only be along the wire. **Wires** can be straight or curved and can also be rough or smooth.

An example of mathematical modelling

This example uses some of the techniques that are covered in this module, so it may be best not to look at the fine detail until you have studied the section on dynamics. However, the setting up stage shows how making assumptions is key to being able to solve real problems.

This example is based on a real situation, but the details have been changed for reasons of confidentiality.

Problem statement
A van, which was travelling down a hill, applied its brakes and skidded into a stationary car. The van left a skid mark of 77 feet before hitting the car.

The solicitors representing the driver of the car want to show that the van was breaking the 30 mph speed limit.

Formulation stage
The following assumptions have been made to allow the problem to be solved. The emphasis here has been on creating a simple model, particularly as the information provided leaves a lot of unknown quantities.

- The van is modelled as a particle.

- The road is modelled as a horizontal surface, because there is no information about the actual slope of the road.

- The speed of the van at impact is assumed to be zero, as there is no information available.

- It is assumed that there are no air resistance forces acting on the car.

- The coefficient of friction between the tyres and the road is assumed to be 0.8, which is a typical value for tyres on a road surface.

- The mass of the van is taken to be m kg as it also is unknown.

- The length of the skid is 23.1 metres, based on 1 foot being 0.3 metres.

Mathematical solution

Figure 1.3 shows the forces acting on the van, when modelled as a particle. R represents the upward reaction force, mg (mass × the acceleration due to gravity) the weight of the van and F shows the friction force.

Fig. 1.3
The forces acting on the van.

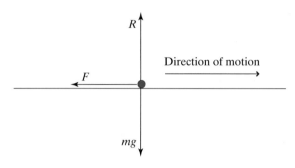

The vertical forces must balance, so $R = mg$

The resultant force will simply be due to the friction force. This force opposes the motion and so the resultant force will be negative:

$$\text{resultant force} = -F$$

But F can be calculated using $F = \mu R$ and since in this case $R = mg$ the resultant force becomes:

$$\text{resultant force} = -\mu R = -\mu mg$$

where μ is the coefficient of friction.

Applying Newton's Second Law gives:

$$ma = -\mu mg$$
$$a = -\mu g$$
$$= -0.8 \times 9.8$$
$$= -7.84 \text{ m s}^{-2}$$

The constant acceleration equation, $v^2 = u^2 + 2as$, can be used with the following values; $v = 0$, $a = -7.84$ and $s = 23.1$

$$v^2 = u^2 + 2as$$
$$0^2 = u^2 + 2 \times (-7.84) \times 23.1$$
$$u = \sqrt{2 \times 7.84 \times 23.1} = 19 \text{ m s}^{-1} \text{ (to 2 sf)}$$

Review and interpretation

The first important task is to convert the speed from m s^{-1} to mph, so that the result can be understood.

$$19 \text{ m s}^{-1} = \frac{19 \times 60^2}{1609} = 43 \text{ mph}$$

While it could be concluded that the van driver was breaking the speed limit, there are a number of criticisms that could be made of this solution. The three main criticisms that could be made are:

Essential notes

Newton's Second Law states that if the resultant force on a body is not zero, that the body will accelerate. The relationship can be expressed mathematically as:

$F = ma$
where:
F = resultant force
m = mass
a = acceleration

Essential notes

The constant acceleration equation $v^2 = u^2 + 2as$ links the following variables:

v = final velocity
u = initial velocity
a = acceleration
s = displacement

- The slope of the road was not taken into account.
- The speed of the van would have been greater than zero when it hit the car.
- The coefficient of friction might not have been equal to 0.8.

A revised model can investigate how these factors could be taken into account.

Revised formulation

The revised assumptions are listed below.

- The van is modelled as a particle.
- The road is modelled as an inclined plane at an angle θ, where $0 \leq \theta \leq 5°$.
- The speed of the van at impact is v, where $0 \leq v \leq 9 \text{ m s}^{-1}$. This allows for an impact speed of up to 20 mph.
- There are no resistance forces acting on the car.
- The coefficient of friction between the tyres and the road is μ, where $0.7 \leq \mu \leq 0.9$.
- The mass of the van is taken to be m kg as it also is unknown.
- The length of the skid is 23.1 metres, based on 1 foot being 0.3 metres.

Revised solution

Figure 1.4 shows the forces acting on the van.

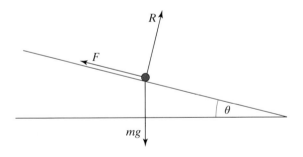

Fig. 1.4
The forces acting on the van.

Resolving perpendicular to the slope gives:

$$R = mg \cos \theta$$

The resultant force can be found by resolving parallel to the slope:

$$
\begin{aligned}
\text{Resultant Force} &= mg \sin \theta - F \\
&= mg \sin \theta - \mu R \\
&= mg \sin \theta - \mu mg \cos\theta \\
&= mg(\sin \theta - \mu \cos\theta)
\end{aligned}
$$

Applying Newton's Second Law gives:

$$
\begin{aligned}
ma &= mg(\sin \theta - \mu \cos \theta) \\
a &= g(\sin \theta - \mu \cos \theta)
\end{aligned}
$$

Essential notes

The weight can be split into two components $mg\cos\theta$ perpendicular to the slope and $mg\sin\theta$ parallel to the slope.

Using the constant acceleration equation $v^2 = u^2 + 2as$, gives:

$$v^2 = u^2 + 2as$$
$$v^2 = u^2 + 2 \times g(\sin\theta - \mu\cos\theta) \times 23.1$$
$$u = \sqrt{v^2 - 453.76(\sin\theta - \mu\cos\theta)}$$

This equation can be used with the ranges of values for v, μ and θ, to produce a range of values for v. The table below shows these values.

v	μ	θ	u (m s^{-1})	U (mph)
0	0.7	0	17.8	40
0	0.7	5	16.6	37
0	0.9	0	20.2	45
0	0.9	5	19.2	43
9	0.7	0	20.0	45
9	0.7	5	18.9	42
9	0.9	0	22.1	49
9	0.9	5	21.2	47

Review and interpretation

From the table it can be seen that the values for the initial speed range are between 37 and 49 mph. These can be presented with much more confidence because all of the potential factors have been considered. The only factor that has been considered and could be criticised is that air resistance has been ignored. This quantity is very much more difficult to take into account. The air resistance would increase the overall force opposing the motion and hence increase the deceleration. This would mean that the initial speeds calculated would be higher and not change the conclusion that the van was breaking the 30 mph speed limit.

Kinematics is the name given to the study of motion. It does not try to explain why motion takes place. This is considered in the **dynamics** section. It does however provide the terms and ideas needed to fully describe and understand motion.

Graphical methods

Kinematics problems can be attempted using a variety of different graphs. The one that is used most often is a velocity-time graph, but you may also meet displacement-time, speed-time and acceleration-time graphs.

Displacement-time graphs

Figure 2.1 shows an example of a **displacement-time graph**. In this case the object moves away from the origin and then returns to it later. While the height of the graph is increasing (and the gradient positive) the object is moving away from the origin. After the peak, the graph is decreasing (and the gradient negative) and the object is moving towards the origin.

The gradient of the graph gives the velocity at that time. The speed of an object tells us how fast it is going, while the velocity gives both the speed and the direction of motion. In straight line kinematics one direction will be positive and the other negative.

Fig. 2.1
A displacement-time graph.

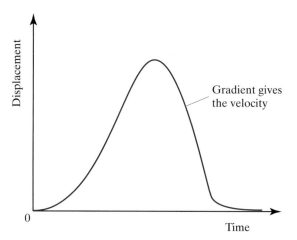

Velocity-time graphs

Figure 2.2 shows a **velocity-time graph**. This is the most useful type of graph as the **acceleration** can be found from the gradient and the displacement from the area under the graph.

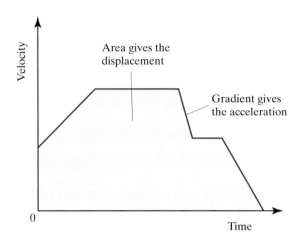

Fig. 2.2
A velocity-time graph.

Speed-time graphs

A **speed-time** graph is very similar to a velocity-time graph, but is different because speed is always positive. One direction, for example to the right, could be defined as positive and the opposite direction, in this case to the left would be negative. If the particle moves towards the left its velocity will be negative. Figure 2.3 shows a speed-time and a velocity-time graph for the same motion.

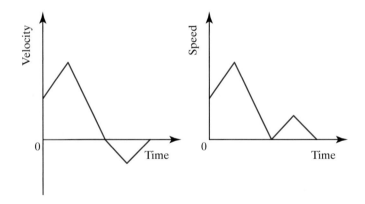

Fig. 2.3
Velocity-time and speed-time graphs for the same motion.

Note that the first part of the graph is the same on both graphs. However, the second part is not the same as the speed is positive while the velocity is negative. The velocity-time graph shows the direction in which the object is moving, but the speed time graph does not do this.

Acceleration-time graphs

Figure 2.4 shows an **acceleration-time graph**. The area under the graph gives the velocity of the moving body.

Fig. 2.4
An acceleration-time graph.

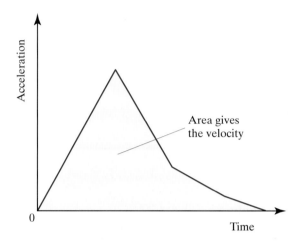

Example

The graph below, in Figure 2.5, is a velocity-time graph for a train moving on a straight length of track.

a) Find the total distance travelled by the train.

b) Find the distance between the train at the end of its journey and its starting point.

Fig. 2.5
Velocity-time graph for the train.

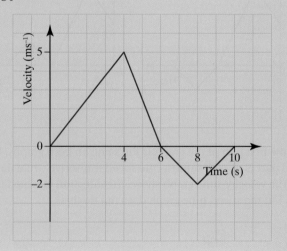

Method notes

The train travels forwards and then backwards. The distance travelled forward is given by the area of the triangle on the left and the distance travelled backwards is given by the area of the triangle on the right.

The total distance is the sum of these two areas.

The distance from the starting point is the difference between the two distances.

Answer

Distance travelled forward (s_1):

$$s_1 = \frac{1}{2} \times 6 \times 5 = 15 \text{ m}$$

Distance travelled backwards (s_2):

$$s_2 = \frac{1}{2} \times 4 \times 2 = 4 \text{ m}$$

a) Total distance travelled = $15 + 4 = 19$ m

b) Distance from starting point = $15 - 4 = 11$ m

Method notes

The velocity of the lift increases by 0.5 m s^{-1} every second, so after 6 seconds it reaches a velocity of 3 m s^{-1}.

Each line on the graph represents one of the three stages of motion. Note that the 3 m s^{-1} found in (a) is the constant speed for the second stage.

Example

A lift, which is initially at rest, rises with a constant acceleration of 0.5 m s^{-2} for 6 seconds. It then travels at a constant speed reached for 8 seconds. It then decelerates uniformly to rest in a further 4 seconds.

a) Calculate the velocity of the lift at the end of the first 6 seconds.

b) Sketch a velocity-time graph for the lift.

c) Find the distance travelled by the lift.

Answer

a) $v = 0.5 \times 6 = 3 \text{ m s}^{-1}$

b)

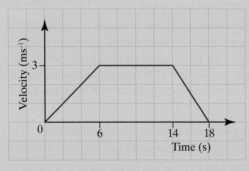

Fig. 2.6
Velocity-time graph for the lift.

c) Distance $= \dfrac{1}{2} \times 6 \times 3 + 8 \times 3 + \dfrac{1}{2} \times 4 \times 3$

$= 9 + 24 + 6$

$= 39$ m

Method notes

The distance travelled is given by the area enclosed by the lines and the time axis. It has been considered as two triangles and a rectangle.

Stop and think 1

A particle moves from a point A along a straight line to the point B, where it remains stationary for 1 second. It then starts to move back along the line towards A. The velocity-time graph for the particle is shown in Figure 2.7.

Fig. 2.7
Velocity-time graph for the particle.

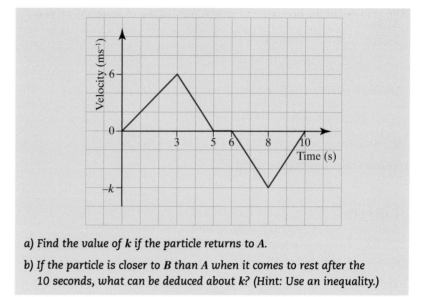

a) *Find the value of k if the particle returns to A.*

b) *If the particle is closer to B than A when it comes to rest after the 10 seconds, what can be deduced about k? (Hint: Use an inequality.)*

Constant acceleration equations

The **constant acceleration equations** are a set of equations that can be used when an object moves with a constant acceleration. The equations connect acceleration, time, initial velocity, final velocity and displacement.

While there are some situations in which the acceleration can reasonably be assumed to be constant, there are others where this is not the case. Within this module, these equations can be used in almost all of the contexts encountered, but in more advanced examples the constant acceleration assumption cannot be made and the equations should not be used.

Deriving the constant acceleration equations

The constant acceleration equations can be derived from the graph shown in Figure 2.8.

Essential notes

In the constant acceleration equations:

u = initial velocity

v = final velocity

t = time taken

a = acceleration

s = displacement

Fig. 2.8
Graph for deriving constant acceleration equations.

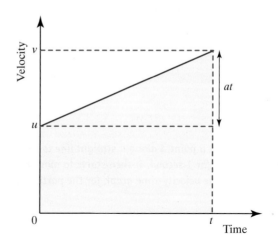

The gradient of the line is a, the acceleration of the moving object. The change in the height of the line on the graph is at which can be added to u to give v. This is the simplest constant acceleration equation:

$$v = u + at$$

The displacement of the moving object is given by the area under the line, which has been shaded in Figure 2.8. The area can be calculated in three ways which leads to three different formulae.

Note that the shaded area is a trapezium. Using the formula for the area of a trapezium $\left(A = \frac{1}{2}(a + b)h \right)$ gives:

$$s = \frac{1}{2}(u + v)t$$

Another way to find the area is to think of the shaded area as being made up of a rectangle and a triangle. This gives the formula:

$$s = ut + \frac{1}{2}at^2$$

A third way to find the area is to start with the rectangle that has height v and to subtract the area of a triangle. This gives the formula:

$$s = vt - \frac{1}{2}at^2$$

The final formula is obtained by eliminating t from two of the above equations. The easiest way to do this is shown below:

Start with one equation:

$$v = u + at$$
$$t = \frac{v - u}{a}$$

Then substitute this into:

$$s = \frac{1}{2}(u + v)t$$
$$s = \frac{1}{2}(u + v) \times \left(\frac{v - u}{a} \right)$$
$$s = \frac{v^2 - u^2}{2a}$$
$$2as = v^2 - u^2$$
$$v^2 = u^2 + 2as$$

Using the constant acceleration equations

When approaching problems that require the use of constant acceleration equations, a good approach is to list the quantities that you know and the one that you need to find. You can then select the equation that contains these quantities.

Exam tips

Make sure that you learn all five constant acceleration equations for the exam.

$$v = u + at$$
$$s = \frac{1}{2}(u + v)t$$
$$s = ut + \frac{1}{2}at^2$$
$$s = vt - \frac{1}{2}at^2$$
$$v^2 = u^2 + 2as$$

Example

A cyclist moves with constant acceleration as he crosses a hatched yellow box at a junction. He moves in a straight line as he moves across the box and travels 20 m. When he enters the box he is travelling at 3 m s^{-1} and when he leaves at 7 m s^{-1}.

a) Find the time that it takes the cyclist to travel the 20 m.

b) Find the acceleration of the cyclist.

Answer

a) $u = 3$, $v = 7$ and $s = 20$, then $t = ?$

$$s = \frac{1}{2}(u + v)t$$

$$20 = \frac{1}{2}(3 + 7)t$$

$$20 = 5t$$

$$t = \frac{20}{5} = 4 \text{ s}$$

b) $u = 3$, $v = 7$ and $s = 20$, then $a = ?$

$$v^2 = u^2 + 2as$$

$$7^2 = 3^2 + 2 \times a \times 20$$

$$49 = 9 + 40a$$

$$a = \frac{49 - 9}{40} = 1 \text{ m s}^{-2}$$

Example

A car is travelling along a straight road when the driver sees a red traffic light ahead. The car is moving at 26 m s^{-1} and is 80 m from the traffic lights when the driver applies the brakes and decelerates at a constant 3 m s^{-2}.

a) Find the time that it takes to reach the traffic lights.

b) Find the speed of the car when it reaches the traffic lights.

c) Find the deceleration that would have brought the car to rest at the traffic lights.

Answer

a) $u = 26$, $a = -3$ and $s = 80$, then $t = ?$

$$s = ut + \frac{1}{2}at^2$$

$$80 = 26t - 1.5t^2$$

$$1.5t^2 - 26t + 80 = 0$$

Solving this quadratic equation with the standard formula gives:

$$t = \frac{26 \pm \sqrt{26^2 - 4 \times 1.5 \times 80}}{2 \times 1.5}$$

$$t = \frac{26 \pm 14}{3} = 4 \text{ or } \frac{40}{3}$$

$$t = 4 \text{ s}$$

b) $u = 26$, $a = -3$ and $s = 80$, then $v = ?$

$$v^2 = u^2 + 2as$$

$$v^2 = 26^2 + 2 \times (-3) \times 80$$

$$= 196$$

$$v = \sqrt{196} = 14 \text{ m s}^{-1}$$

c) $u = 26$, $v = 0$ and $s = 80$, then $a = ?$

$$v^2 = u^2 + 2as$$

$$0^2 = 26^2 + 2 \times a \times 80$$

$$0 = 676 + 160a$$

$$a = -\frac{676}{160} = -4.225 \text{ m s}^{-2}$$

Stop and think 2

A particle is moving on the line AB shown in Figure 2.9. The direction AB is taken as positive as shown by the diagram.

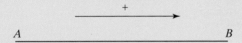

State whether the velocity and the acceleration of the particle are positive or negative in each case below.

a) The particle is moving from A towards B and slowing down.

b) The particle is moving from B towards A and speeding up.

c) The particle is moving from B towards A and slowing down.

Fig. 2.9
Definition of direction.

Vertical motion under gravity

The constant acceleration equations can be used to solve problems involving vertical motion. In these problems it is very important to think carefully about the signs of the velocity and the acceleration.

If the initial motion is upwards then it is good to take this direction as positive and use the acceleration as $-g$ or -9.8 m s^{-2}.

If the object is falling from rest or has projected downwards, then use $+g$ or 9.8 m s^{-2}.

Using these values makes the assumption that the weight is the only force acting on the moving object and that there are no resistance forces.

Method notes

When the ball reaches its maximum height the velocity of the ball is zero ($v = 0$). This gives a condition that allows the constant acceleration equation to be used. This part is completed by adding the initial height of the ball.

Time to the top, the time taken to reach the maximum height and the time taken for the ball to fall from this point to the ground are needed. Once found they can be added together.

The final answer is given to 2 significant figures as the value of **g** was used correct to 2 significant figures.

Example

A ball is thrown vertically upwards with an initial speed of 11.2 m s^{-1} from a height of 0.8 m.

a) Find the maximum height of the ball above the ground.

b) Find the time it takes for the ball to return to the ground after it has been thrown.

Answer

a) $v^2 = u^2 + 2as$ with $v = 0, u = 11.2$ and $a = -9.8$

$0^2 = 11.2^2 + 2 \times (-9.8) \times s$

$0 = 11.2^2 - 19.6s$

$s = \dfrac{11.2^2}{19.6} = 6.4$ m

$h = 6.4 + 0.8 = 7.2$ m

b) Time up

$v = u + at$ with $v = 0, u = 11.2$ *and* $a = -9.8$

$0 = 11.2 - 9.8t$

$t = \dfrac{11.2}{9.8} = 1.14$ s

Time down

$s = ut + \dfrac{1}{2}at^2$ with $s = 7.2, u = 0$ and $a = 9.8$

$7.2 = 4.9t^2$

$t = \sqrt{\dfrac{7.2}{4.9}} = 1.21$ s

Total time $= 1.14 + 1.21 = 2.35$ s (rounded to 2.4 s)

Example

A ball is released from rest at a height of 15 m above ground level. Assume that there is no resistance to its motion.

a) Find the time it takes the ball to reach the ground.

b) Find the speed of the ball when it hits the ground.

Answer

a) $s = ut + \dfrac{1}{2}at^2$ with $s = 15, u = 0$ and $a = 9.8$

$15 = 0t + \dfrac{1}{2} \times 9.8t^2$

$15 = 4.9t^2$

$t = \sqrt{\dfrac{15}{4.9}} = 1.7$ s (to 2 sf)

b) $v^2 = u^2 + 2as$ with $u = 0, a = 9.8$ and $s = 15$

$v^2 = 0^2 + 2 \times 9.8 \times 15$

$v = \sqrt{2 \times 9.8 \times 15} = 17$ m s^{-1} (to 2 sf)

Stop and think 3

A particle falls from rest from a height of 2h metres and after T seconds hits the ground at a velocity V m s^{-1}.

After $\dfrac{T}{2}$ seconds it has fallen a distance x metres and has velocity v m s^{-1}.

Which of the following are true?

$v = \dfrac{V}{2}$

$v > \dfrac{V}{2}$

$v < \dfrac{V}{2}$

$x = h$

$x > h$

$x < h$

Stop and think answers

1 a) $k = 7.5$ b) $k < 3.75$

2 a) Velocity is positive. Acceleration is negative.

 b) Velocity is negative. Acceleration is negative.

 c) Velocity is negative. Acceleration is positive.

3 First consider the velocity:

 After falling for T seconds:

  $V = 0 + 9.8T$

  $V = 9.8T$

 After falling for $\dfrac{T}{2}$ seconds:

  $v = 0 + 9.8 \times \dfrac{T}{2}$

  $v = 4.9T = \dfrac{V}{2}$

  $\therefore v = \dfrac{V}{2}$ is tue.

 Now consider the distance fallen:

 After falling for T seconds:

  $2h = \dfrac{1}{2} \times 9.8T^2$

  $h = \dfrac{9.8T^2}{4}$

 After falling for $\dfrac{T}{2}$ seconds:

  $x = \dfrac{1}{2} \times 9.8\left(\dfrac{T}{2}\right)^2$

  $x = \dfrac{9.8T^2}{8} = \dfrac{1}{2}h$

  $\therefore x < h$ is true

Types of force

A number of different types of force will be encountered in mechanics contexts. The main ones will be considered here.

Weight

The weight is the gravitational attraction that the Earth exerts on objects. This force must be considered when working with objects on or close to the surface of the Earth. The Earth also exerts a gravitational attraction on the Moon, which is why it remains in orbit around the Earth. In this module, only objects on or close to the Earth will be considered. The weight is directed towards the centre of the Earth, but in most problems this can simply be considered as vertically downwards.

The weight should be represented on a diagram by an arrow that is directed vertically downwards and starts at the centre of mass of the body being considered. The centre of mass of the body is considered in later modules, but for uniform objects, it will be located at the centre of the object. Figure 3.1 shows some bodies and how the weight force is represented by a downward arrow.

 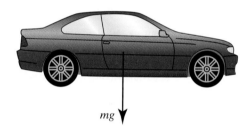

For the football the centre of mass is clearly at the centre of the ball. For the car it is more difficult to know where the centre of mass is, but for the purpose of a diagram it is enough to show that it acts somewhere inside the car.

If an object is modelled as a particle, then the arrow to represent the weight is easy to draw in as it must start at the particle.

Essential notes

The weight acts towards the centre of the Earth and has magnitude *mg* where *g* is the acceleration due to gravity.

On Earth: $g = 9.8 \text{ m s}^{-2}$.

Fig. 3.1
Arrows being used to represent the weight.

Essential notes

When calculating the weight of an object, the mass must be in kg.

1 kg = 1000 grams

1 tonne = 1000 kg

Method notes

Use *mg* to find the weight, first converting the mass to kg if necessary, as in (b) and (c).

Example

Calculate the magnitude of the weight of:

a) a car of mass 1100 kg

b) a ball of mass 150 grams

c) a lorry of mass 30 tonnes.

Answer

a) Weight $= 1100 \times 9.8 = 10\,780$ N

b) Weight $= 0.15 \times 9.8 = 1.47$ N

c) Weight $= 30\,000 \times 9.8 = 294\,000$ N

Normal reaction

Whenever two bodies are in contact they will exert forces on each other. Often these forces act to so that the overall force on an object is zero. In this state the forces on a particle are said to be in **equilibrium**.

A simple case is when an object is placed on a horizontal surface, such as a mug placed on a desk. The weight of the mug pulls down on the mug, but the desk pushes upwards on the mug to balance the downward force. Figure 3.2 shows the forces acting on the mug. The normal reaction, R, is represented by an arrow which begins at the point of contact. In this case $R = mg$, so that the forces balance.

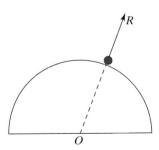

Essential notes

The units of all forces are newtons (N). As weight is a force it has units of newtons. Note that weight is often confused with mass in everyday life, but mass has units of kg. Be careful not to confuse these two different quantities.

Fig. 3.2
The forces acting on the mug.

The word **normal** is used to describe the direction of the reaction force and has the same meaning as in the core modules where tangents and normals are considered. When an object is on a plane surface the normal reaction force will be perpendicular to the surface.

Even if the surface is not horizontal, the normal reaction force will act in a direction that is perpendicular to the surface, as shown in Figure 3.3, where the particle is on an inclined plane. In this case the normal reaction force is not equal to the weight.

If a particle is placed on a curved surface, then the normal reaction acts along a normal to the surface. Figure 3.4 shows the direction of the reaction force of a particle placed on a hemisphere, with centre O. The normal reaction force acts along an extension of the radius.

Fig.3.3
The normal reaction force on a particle on an inclined plane.

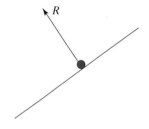

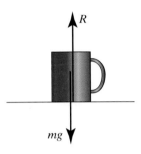

Fig.3.4
The normal reaction on a particle on a hemisphere.

Often when an object is placed on a horizontal surface the normal reaction force is equal to the weight of the object. However, is important to realise that the normal reaction is not always equal to the weight of the body.

Fig. 3.5
A case where the normal reaction is not equal to the weight.

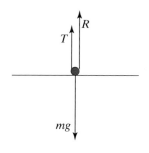

Fig. 3.6
The forces acting on a particle on an inclined plane.

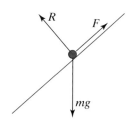

Fig. 3.7
The forces acting on a particle on a hemisphere.

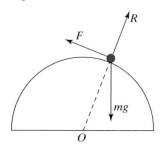

Essential notes

The friction law states that:

$F \leq \mu R$ where:

F = friction

μ = coefficient of friction

R = normal reaction

Fig. 3.8
Pushing the calculator face up and face down.

Figure 3.5 shows a particle, which is at rest on a horizontal surface and attached to an elastic string. This string exerts an upward force on the particle, as does as the normal reaction and, together, these forces balance the weight to produce equilibrium.

In this case:

$R + T = mg$
$R = mg - T$

> ### Stop and think 1
>
> In which of the following situations is the normal reaction force equal to the weight?
>
> a) A particle which is at rest at the top of a hemisphere.
>
> b) A book at rest on an inclined plane.
>
> c) A sledge being pulled on a horizontal surface by a rope which is angled above the horizontal.
>
> d) A can at rest on a horizontal surface.

Friction
Friction acts when objects are in contact, and opposes potential motion. Figure 3.6 shows the forces on a particle that has been placed on an inclined plane. The particle will tend to slide down the plane, but if the friction force, F, is big enough it will remain at rest.

Consider again the particle on the hemisphere. Figure 3.7 shows all the forces acting on the particle. The particle will tend to slide off the hemisphere unless the friction acting along the tangent as shown is strong enough to prevent this.

The friction law
There is a simple law which determines the maximum amount of friction in any situation. This depends on the coefficient of friction, μ, and the magnitude of the normal reaction, R.

The coefficient of friction depends on the two materials that are in contact. Between materials like rubber and wood there are high coefficients of friction, while icy surfaces would have low coefficients of friction.

Friction experiment
A calculator with small rubber feet on one side is ideal for this experiment. Place the calculator on a desk and try to push it with your finger, as shown in Figure 3.8. Then turn the calculator over so that the buttons are face down and try to push again.

You will feel a significant difference in the friction. This is because the coefficients of friction are very different for each side of the calculator.

Calculating the maximum friction

The maximum friction available is given by $F = \mu R$.

- If an object is sliding friction will always take this maximum value.

- If an object is at rest there will be just enough friction to prevent sliding and this will be less than the maximum value of μR.

Example

A book, of mass 2 kg, is placed on a rough horizontal surface. The coefficient of friction between the book and the table is 0.75.

A horizontal force of magnitude P N is applied to the book to push it forward.

Find the magnitude of the friction force and describe what happens to the book if:

a) $P = 10$

b) $P = 14.7$

c) $P = 20$

Answer

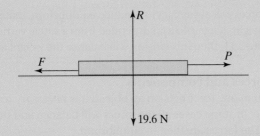

Vertically: $R = mg$

$$= 2 \times 9.8$$

$$= 19.6$$

Horizontally: $F \leq \mu R$

$$F \leq 0.75 \times 19.6$$

$$F \leq 14.7 \text{ N}$$

a) When $P = 10$, then $F = 10$ and the book remains at rest as the forces are balanced.

b) When $P = 14.7$, then $F = 14.7$ and the book remains at rest on the point of sliding as the forces are just balanced.

c) When $P = 20$, then $F = 14.7$ and the book begins to slide on the surface as the friction does not balance the 20 N force.

Fig. 3.9
The forces on the book.

Method notes

First draw a force diagram, as shown in Figure 3.9.

Then find the normal reaction force and calculate the maximum possible amount of friction.

If the applied force, P, is less than or equal to the maximum friction the book will remain at rest.

If the applied force is equal to the maximum friction, the book is said to be on the point of sliding.

If the applied force exceeds the maximum friction, the book will begin to slide.

Stop and think 2

A box, of mass 2 kg, is on a rough horizontal surface. When a horizontal force of 14 N is applied to the box it is on the point of sliding. A particle, of mass 3 kg, is placed in the box. What horizontal force must now be applied for the box to be on the point of sliding?

Tension

When strings are involved in a situation, a tension force will be present and it will act along the direction of the string. Figure 3.10 shows the forces acting on a particle suspended from two strings.

Fig. 3.10
The forces on a particle suspended by two strings.

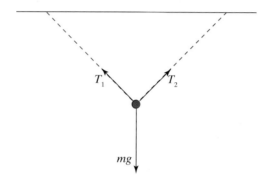

Fig. 3.11
The forces on a particle suspended by a single string.

The tension forces will always act to produce **equilibrium**, that is they balance the weight. If a particle is suspended from a single vertical string, as shown in Figure 3.11, then this tension will balance the weight and $T = mg$.

Resultant forces and components

When more than one force acts on an object, it is necessary to consider the resultant force. In some cases all the forces will balance and the resultant force is zero. The forces are then in equilibrium.

When the forces are not in equilibrium, the forces will be equivalent to a single resultant force. With forces that act only in the horizontal and vertical this is very easy to find.

Figure 3.12 shows four forces. If these are to be in equilibrium, then $P = 17$ so that the vertical forces balance and $Q = 15$, so that the horizontal forces balance.

However, if $Q = 15$ and $P = 20$, then the vertical forces would not balance and there would be an upward resultant force of $20 - 17 = 3$ N.

Fig. 3.12
Four forces in equilibrium.

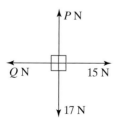

Example

In each case shown below in Figure 3.13, describe the resultant of the four forces acting.

a)

b)

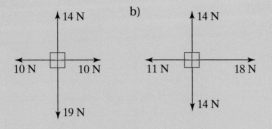

Fig. 3.13
Finding the resultant force.

Method notes

Note that in each case either the horizontal or vertical forces balance.

The resultant can be found by looking at the difference of the forces in the other direction.

Answer

a) The resultant is 19 − 14 = 5 N vertically downwards.

b) The resultant is 18 − 11 = 7 N horizontally to the right.

Using components of forces

Forces that are either vertical or horizontal are very easy to deal with, but many forces act in other directions. In this case it is possible to split the forces in to two perpendicular **components**. These two directions are normally either horizontal and vertical or parallel and perpendicular to a slope. Sometimes if a problem involves two perpendicular forces a quick solution can be found by resolving in the directions of these two forces.

Figure 3.14 shows two examples of resolving forces. In (a) the force has been resolved into horizontal components and in (b) the forces has been resolved into components parallel and perpendicular to the dotted line.

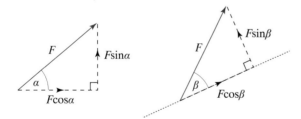

Fig. 3.14
Resolving forces.

Example

Figure 3.15 shows three forces which act in a vertical plane. Show that the resultant of these forces is horizontal and find the magnitude of the resultant force.

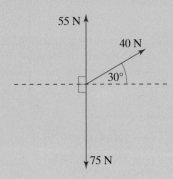

Answer

Vertically:

$$55 + 40 \sin 30° - 75 = 55 + 20 - 75 = 0$$

The vertical components give a sum of zero and so the resultant force must be horizontal.

Horizontally:

$$40 \cos 30° = 34.6 \text{ N (to 3sf)}$$

The resultant force is 34.6 N to the right.

Example

A child on a bike freewheels down a hill. As the child moves, a resistance force of 30 N acts to oppose the motion. Assume the child and bike can be modelled as a particle of mass 40 kg and that the hill can be modelled as a slope inclined at 5° to the horizontal.

Find the magnitude of the resultant force on the child and bike.

Answer

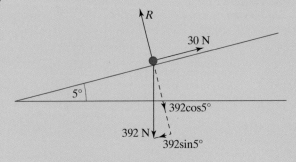

$$
\begin{aligned}
\text{Resultant Force} &= 392 \sin 5° - 30 \\
&= 4.2 \text{ N (to 2sf)}
\end{aligned}
$$

Newton's Laws

Newton's First Law

Newton's First Law is very important. If the resultant or overall force on an object is zero, then it will either remain at rest or continue to move with a constant velocity. The converse is also true, if a body is moving a constant velocity, then the resultant force on the body will be zero.

Consider a car travelling at its maximum speed along a straight road. In this case the velocity is constant and so the resultant force on the car must be zero. Figure 3.17 shows the forces on the car.

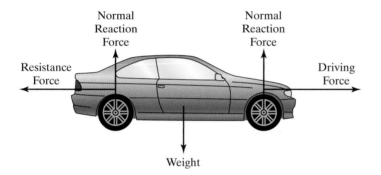

Clearly the vertical forces must balance. The horizontal forces must also balance. If they were not balanced, then either the car would gain speed if the driving force were greater than the resistance force, or the car would lose speed if the resistance force were greater than the driving force.

> ### Stop and think 3
>
> *In which of the following cases is the resultant force zero? Explain why.*
>
> *a) A parachutist falling vertically at a constant speed.*
>
> *b) A car travelling round a roundabout at a constant speed.*

Newton's Second Law

Newton's Second Law describes what happens when the resultant force on a body or object is not zero. Applying Newton's Second Law is one of the most commonly used and important ideas in mechanics. In order to be systematic and avoid errors the following procedure is recommended:

1. Draw a diagram to show the forces.

2. Find the resultant force.

3. Apply Newton's Second Law.

4. Solve the resulting equation(s).

Essential notes

Newton's First Law states that a body will remain at rest or move with a constant velocity unless acted on by a resultant force.

Fig. 3.17
The forces on a car at its maximum speed.

Essential notes

Newton's Second Law states that if the resultant force on a body is not zero, that the body will accelerate. The relationship can be expressed mathematically as:

$F = ma$

where:

F = resultant Force

m = mass

a = acceleration

Example

A cyclist and cycle are modelled as a particle, of mass 80 kg, travelling on a horizontal surface. A horizontal forward force of 120 N causes the cyclist to accelerate at 0.8 m s^{-2}. A constant resistance force of magnitude P N also acts on the cyclist. Find P.

Answer

Fig. 3.18
The forces and acceleration for the cyclist.

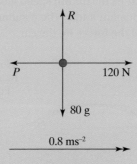

Method notes

First a diagram, Figure 3.18, is drawn to show the forces on the cyclist.

The resultant of the horizontal forces is then found. (Note that the vertical forces balance each other.)

Newton's Second Law is then applied and the resulting equation solved to find P.

resultant force $= 120 - P$
$$F = ma$$
$$120 - P = 80 \times 0.8$$
$$120 - P = 64$$
$$P = 120 - 64$$
$$= 56$$

Example

A woman of mass 60 kg is standing in a lift, which is rising, but slowing down, with a deceleration of 0.05 m s^{-2}. Find the magnitude of the normal reaction exerted by the lift floor on the woman.

Answer

Fig. 3.19
The acceleration and forces for the woman in the lift.

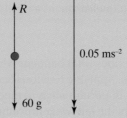

resultant force $= R - 60g$
$$F = ma$$
$$R - 60g = 60 \times (-0.05)$$
$$R - 588 = -3$$
$$R = 588 - 3$$
$$= 585 \text{ N}$$

Newton's Third Law

Newton's Third Law is often misunderstood because those using it do not realise that the opposite reaction has to be the same type of force as the action. Both forces in the action-reaction pair must be of the same type.

For example if a book is placed on a desk, then the book pushes down on the desk and exerts a downward normal reaction force on the desk. The desk also pushes up on the book and exerts an upward normal reaction force on the book, as shown in Figure 3.20. These two forces form an action-reaction pair.

It is important to note that this pair of forces does not involve the weight of the book. It is the two normal reaction forces that make the pair.

This leads to the question: 'What is the opposite reaction to the weight force?'

Any object on Earth experiences a weight force due to the gravitational attraction of the Earth on the object. The object also exerts a force on the Earth, as shown in Figure 3.21. The Earth exerts a force on the object towards the centre of the Earth and the object exerts an equal but opposite force on the Earth. This force acts away from the centre of the Earth. (The magnitude of this force is significant for the object, but not the Earth which has a much greater mass.)

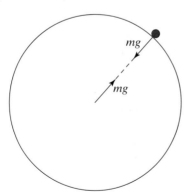

Method notes

First draw a diagram, Figure 3.19, to show the forces on the woman and her acceleration. Note that the force exerted by the floor is the only upward force acting on the woman.

In this case $a = -0.05$ m s^{-2} as the woman is decelerating.

The resultant of the vertical forces is then found.

Newton's Second Law is then applied and the resulting equation solved to find R.

Fig. 3.20
The upward normal reaction force on the book and the downward normal reaction force on the desk.

Essential notes

Newton's Third Law states that for every action (force) there is an equal and opposite reaction.

Fig. 3.21
The action and reaction pair due to gravity.

Stop and think 4

A box is placed on a rough table and pushed so that it moves horizontally. As the box moves it experiences friction. Describe the Newton's Third Law reaction to this friction force.

Simple experiment

Find a calculator, or other object, that has small rubber feet.

Place the calculator on a sheet of paper on a desk. Exert a horizontal force on the calculator, so that it moves. Explain why the paper moves too. What forces are acting on the paper?

Motion on an inclined plane

Many examples in mechanics involve motion on an inclined plane or slope. To solve these problems the weight of the object, and sometimes other forces acting need to be resolved parallel and perpendicular to the plane. In this context some forces, including friction and the normal reaction act parallel and perpendicular to the slope and working parallel and perpendicular to the slope makes them very easy to deal with.

Figure 3.22 shows the forces acting on a particle on a rough slope and the two components of the weight force. The friction force acts up the slope as the particle will tend to slide down the slope.

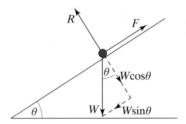

Once the force diagram has been drawn, two equations can be written down:

- One by using the components of the forces perpendicular to the slope, which will usually be in equilibrium.

- A second by applying Newton's Second Law to the resultant of the components of the forces parallel to the slope.

Essential notes

The angle between the weight and its component perpendicular to the slope is the same as the angle between the slope and the horizontal. Note the two angles marked θ in Figure 3.22.

Fig. 3.22
The forces acting on a particle on an inclined plane.

Exam tips

When showing the components of a force on a diagram, use a different notation, such as the dashed lines shown in Figure 3.22.

Example

A child, of mass 30 kg, is on a slide inclined at 40° to the horizontal. The coefficient of friction between the child and the slide is 0.6. Assume that no air resistance forces act on the child.

a) Find the acceleration of the child down the slide.

b) Find the speed of the child when she has moved 3 m down the slide. Assume that she starts at rest.

Fig. 3.23
The forces on the child.

Answer

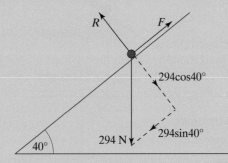

a) Perpendicular to the slope. The reaction force is equal to the component of the weight perpendicular to the slope:

$R = 294 \cos 40°$

Parallel to the slope. The resultant force is the component of the weight parallel to the slope minus the friction:

Resultant Force $= 294 \sin 40° - F$

Applying Newton's Second Law down the slope:

$$294 \sin 40° - F = 30a$$
$$294 \sin 40° - \mu R = 30a$$
$$294 \sin 40° - 0.6 \times 294 \cos 40° = 30a$$
$$a = \frac{294 \sin 40° - 0.6 \times 294 \cos 40°}{30} = 1.794 \ldots = 1.8 \text{ m s}^{-2} \text{ (to 2 sf)}$$

b) $v^2 = u^2 + 2as$

$v^2 = 0^2 + 2 \times 1.79 \times 3$

$v = \sqrt{2 \times 1.79 \times 3} = 3.28 \ldots = 3.3 \text{ m s}^{-1} \text{ (to 2 sf)}$

Method notes

First draw a force diagram and identify the components of the weight, as in Figure 3.23.

Find the normal reaction by considering the components of the forces perpendicular to the slope. Note that this will be needed to find the friction force.

Apply Newton's Second Law parallel to the slope to find the acceleration.

Once the acceleration is known, it can be used in a constant acceleration equation to find the final speed of the child.

Method notes

The force diagram is shown in Figure 3.24.

Considering the forces perpendicular to the slope allows the normal reaction to be found.

Applying Newton's Second Law parallel to the slope allows the friction to be found.

These can then be used with $F = \mu R$ to find the coefficient of friction.

Fig. 3.24
The forces on the particle.

Exam tips

The exact expressions for F and R have been used in the equation $F = \mu R$ to ensure that no rounding errors are introduced.

Example

A particle of mass 6 kg is placed on a rough slope inclined at an angle of 20° to the horizontal. When the particle is released it accelerates down the slope at 0.2 m s⁻². Find the coefficient of friction between the slope and the particle.

Answer

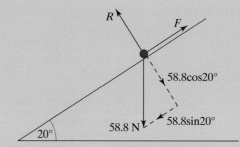

Perpendicular to the slope:

$$R = 58.8 \cos 20°$$

Parallel to the slope:

$$\text{Resultant Force} = 58.8 \sin 20° - F$$

Applying Newton's Second Law down the slope:

$$58.8 \sin 20° - F = 6 \times 0.2$$
$$F = 58.8 \sin 20° - 6 \times 0.2$$
$$F = \mu R$$
$$58.8 \sin 20° - 6 \times 0.2 = \mu(58.8 \cos 20°)$$
$$\mu = \frac{58.8 \sin 20° - 6 \times 0.2}{58.8 \cos 20°} = 0.342\ldots = 0.34 \text{ (to 2 sf)}$$

Example

A child of mass 45 kg sits on a sledge, which slides down a rough slope inclined at 15° to the horizontal. Assume that a constant air resistance force, of magnitude 20 N, acts on the child. Model the child and the sledge as a particle of mass 45 kg. The coefficient of friction between the sledge and the particle is 0.2. Find the acceleration of the child and sledge.

Answer

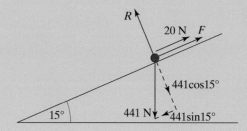

Fig. 3.25
The forces on the sledge.

Perpendicular to the slope:

$R = 441\cos 15°$

Parallel to the slope:

Resultant Force $= 441 \sin 15° - 20 - F$

Applying Newton's Second Law down the slope:

$$441 \sin 15° - 20 - F = 45a$$
$$441 \sin 15° - 20 - \mu R = 45a$$
$$441 \sin 15° - 20 - 0.2 \times 441 \cos 15° = 45a$$
$$a = \frac{441 \sin 15° - 20 - 0.2 \times 441 \cos 15°}{45} = 0.198\ldots = 0.20\,\text{m s}^{-2}\ \text{(to 2 sf)}$$

Method notes

Start by drawing a force diagram as shown in Figure 3.25.

The normal reaction must be found first so that this can used to find the friction force.

When considering the components of the force parallel to the slope, the resultant force down the slope should be considered.

Example

A car, of mass 1200 kg, drives up a slope inclined at an angle of 5° to the horizontal, accelerating at 0.4 m s^{-2}. A driving force of magnitude P N acts up the slope and a resistance force of 400 N acts down the slope. Find P.

Answer

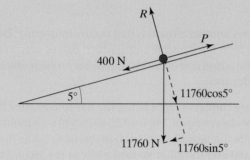

Fig. 3.26
The forces on the car.

Parallel to the slope:

Resultant Force $= P - 400 - 11760 \sin 5°$

Applying Newton's Second Law up the slope:

$$P - 400 - 11760 \sin 5° = 1200 \times 0.4$$
$$P = 1200 \times 0.4 + 400 + 11760 \sin 5°$$
$$= 1904.9\ldots = 1900\,\text{N (to 2 sf)}$$

Method notes

Begin with a force diagram, as shown in figure 3.26.

In this case, as the car is accelerating up the slope, the resultant force up the slope is used.

As friction is not involved there is no need to find the normal reaction force.

Essential notes

When dealing with connected particles, separate the systems and apply Newton's Second Law to each particle.

Connected particles

There are many instances when motion involves two different bodies or objects that are connected in some way. For example a car attached to a trailer or two objects that are joined by a string. The important principle in solving all of these problems is to recognise that there are forces present which have the same magnitude, but that act in different directions. It is very important to separate the two objects and consider the forces acting on each one. In all of the examples below, the two objects have been separated, so that the forces with the same magnitude can be seen easily.

Figure 3.27 shows a car and a caravan. The car exerts a forward force of T N on the caravan and the caravan exerts a backward force of T N on the car.

Fig. 3.27
The car and the caravan.

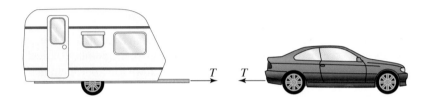

Fig. 3.28
Two objects connected by a string.

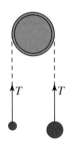

Figure 3.28 shows another situation that occurs frequently. Two objects are connected by a light inextensible string which passes over a smooth peg. In this situation the string exerts the same tension on each object and acts upwards in both cases.

When objects are connected by a string the tensions exerted by each end of the string are always equal. Figure 3.29 shows two situations where this is true, but where the tensions are not both vertical. In the first case one force is horizontal and the other vertical while in the second case one force acts at an angle.

Fig. 3.29
Equal tensions act on two bodies connected by a string.

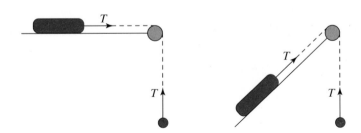

Example

Two particles of masses 3 kg and 2 kg are connected by a light inextensible string that passes over a smooth peg. The particles are released from rest and begin to accelerate.

a) Find the acceleration of the particles

b) Find the tension in the string, giving your answer in terms of g.

Answer

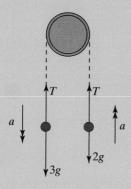

Fig. 3.30
The forces and accelerations.

a) For the 2 kg particle

$$\text{Resultant Force} = T - 2g$$
$$F = ma$$
$$T - 2g = 2a \qquad (1)$$

For the 3 kg particle

$$\text{Resultant Force} = 3g - T$$
$$F = ma$$
$$3g - T = 3a \qquad (2)$$

Adding equations (1) and (2) gives:

$$g = 5a$$
$$a = \frac{g}{5}\,\text{m s}^{-2}$$

b) Substituting this value for the acceleration into the equation for the 2 kg particle gives:

$$T - 2g = 2 \times \frac{g}{5}$$
$$T = \frac{2g}{5} + 2g$$
$$= \frac{12g}{5}\,\text{N}$$

Method notes

Figure 3.30 shows the forces acting on each particle.

For each particle, find the resultant force and apply Newton's Second Law, so that you get two equations.

In this case the two equations have been added together to eliminate T.

Substitute the value for a into one of the two equations to find the tension.

Exam tips

Draw separate force diagrams for each object involved.

Example

A box, of mass 5 kg, is on a rough horizontal surface. The coefficient of friction between the box and the surface is μ. The box is connected by a light inextensible string to a sphere, of mass 4 kg. The string passes over a light pulley, as shown in Figure 3.31.

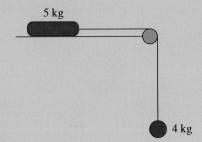

Fig. 3.31
The box and the sphere.

When released from rest, with the string taut, the box and sphere accelerate at 0.9 m s^{-2}. Find μ.

Answer

Fig. 3.32
The forces on the box and the sphere.

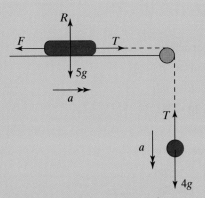

For the 4 kg sphere Newton's Second Law gives:

$$4g - T = 4 \times 0.9$$
$$T = 4g - 3.6$$
$$= 35.6 \text{ N}$$

For the 5 kg box:

$$T - F = 5 \times 0.9$$
$$35.6 - F = 4.5 \quad \text{and} \quad R = 5g = 49 \text{ N}$$
$$F = 35.6 - 4.5$$
$$= 31.1 \text{N}$$
$$F = \mu R$$
$$31.1 = \mu \times 49$$
$$\mu = \frac{31.1}{49} = 0.634\ldots = 0.63 \text{ (to 2 sf)}$$

Method notes

Note that on the diagram in Figure 3.32 each end of the string exerts forces of the same magnitudes, but in different directions.

Start with the sphere and use Newton's Second Law to find the tension in the string.

The normal reaction and friction forces on the box can then be found and used to determine the values of μ.

Example

A van of mass 2000 kg tows a trailer of mass 1200 kg. As they move along a straight horizontal road a resistance force of 400 N acts on the van and a resistance force of 120 N acts on the trailer. A horizontal forward driving force of magnitude P N acts on the van. The van and trailer accelerate at 1.2 m s^{-2}.

a) Find P.

b) Find the magnitude of the force exerted by the trailer on the van.

c) State the magnitude of the force that the van exerts on the trailer.

Answer

a)

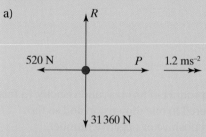

Fig. 3.33
The forces on the van and trailer when considered as a single particle.

Resultant Force $= P - 520$

Applying Newton's Second Law to van and trailer together gives:

$P - 520 = 3200 \times 1.2$

$P = 3200 \times 1.2 + 520$

$= 4360$ N

b)

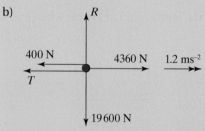

Resultant Force $= 4360 - 400 - T$

Applying Newton's Second Law to the van only gives:

$4360 - 400 - T = 2000 \times 1.2$

$T = 4360 - 400 - 2000 \times 1.2$

$= 1560$ N

c) The van exerts a force of 1560 N on the trailer.

Method notes

In (a) the van and trailer are considered as a single particle of mass 3200 kg. The total resistance force on this is 520 N as shown in the diagram. Figure 3.33 shows the forces acting in this case. Applying Newton's Second Law leads to the forward driving force.

In (b), either the van or trailer has to be considered alone. In this case this has been done with the van, as shown in Figure 3.34. Again applying Newton's Second Law leads to the required force.

Part (c) requires the fact that the van and trailer exert forces of the same magnitude on each other.

Fig. 3.34
The forces exerted on the van alone.

Stop and think 5

Two identical particles, of mass m kg and 2m kg, are on a smooth horizontal surface and connected by a light inextensible string which passes round a smooth light pulley. The string and pulley are also on the smooth horizontal surface. A force is applied to the pulley. This system is shown in Figure 3.35.

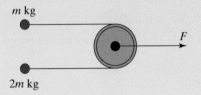

Fig. 3.35
The particles and the pulley. What happens to the particles when the force is applied?

Momentum and impulse

Momentum

The momentum of an object is the product of its mass and velocity. In this module, momentum is only considered in one dimension and so the momentum can be represented as a positive or negative number, depending on the direction of the velocity.

The units of momentum are the product of the units of mass and velocity, which can be written kg m s^{-1}. This is quite a cumbersome way to write the units. An alternative is to use N s (newton seconds) which comes from the link between momentum and impulse that you will see in the next section.

Essential notes

momentum = mass × velocity

Example
A car has mass 1200 kg. Calculate the momentum of the car when it is travelling:

a) Forwards at 20 m s^{-1},

b) Backwards at 0.5 m s^{-1}.

Answer
a) Momentum $= 1200 \times 20$
$\qquad\qquad\quad = 24000$ N s

b) Momentum $= 1200 \times (-0.5)$
$\qquad\qquad\quad = -600$ N s

Method notes

The forward direction has been defined as the positive direction, so that in (a) the velocity is 20 and in (b) when the car is moving backwards the velocity is −0.5.

Impulse
Impulse is defined as the change in the momentum of an object. This definition is usually called the impulse-momentum principle. Impulse is very useful when considering sudden changes in motion, for example in a ball when someone hits it with a bat.

impulse = final momentum – initial momentum

Using the standard notation, with I to represent impulse:

$I = mv - mu$

It is possible to connect impulse, time and force, so that in situations like the ball being hit by a bat where the time is very small and hard to measure, the impulse can be used rather than force and time. In the case where the force and hence the acceleration are assumed to be constant, the constant acceleration equation $v = u + at$ and Newton's Second Law can be used to provide a link. Substituting for v and then using $F = ma$ gives:

$$I = mv - mu$$
$$= m(u + at) - mu$$
$$= mu - mat - mu$$
$$= mat$$
$$= Ft$$

As the impulse is defined as the change in momentum, it must have the same units as momentum. However, the fact that it can also be defined as the product of force and time means that it will have the units N s. This provides an easy way to write both the units on momentum and of impulse.

When fielders catch a fast moving cricket ball they put their hands round it but move them back with the ball. This means that they stop the ball more slowly, that is they extend the time that it takes for the ball to stop. The impulse required to stop the ball is fixed as it depends on the mass and the speed of the ball. However, as $I = Ft$, extending the time reduces the force required and so makes it easier, and less painful, for the fielder to catch the ball.

Example
A ball of mass 0.05 kg is travelling horizontally at 3.2 m s^{-1} when it hits a vertical wall and rebounds at 2.8 m s^{-1}.

Find the magnitude of the impulse on the ball.

Answer
$u = 3.2$, $v = -2.8$ and $m = 0.05$

$$I = mv - mu$$
$$= 0.05 \times (-2.8) - 0.05 \times 3.2$$
$$= -0.3 \text{ Ns}$$

Magnitude of the impulse = 0.3 N s

Essential notes
Remember:

v = final velocity

u = initial velocity

Exam tips
Learn the formulae:

$I = mv - mu$

$I = Ft$

Exam tips
Be very careful with the signs in questions like this.

Method notes
In the question, the most important thing to realise is that one velocity must be positive and the other negative. In this case the initial velocity is taken to be in the positive direction.

Method notes

In (a) the force and time can be used to find the impulse.

Once known, the impulse can be used to find the final velocity.

Example

A toy car has mass 0.25 kg. A horizontal force of 0.8 N is applied to the car for 1.5 seconds.

a) Find the impulse given to the car by this force.

b) Hence find the final speed of the car if was initially at rest.

Answer

a) $I = Ft$
$$= 0.8 \times 1.5$$
$$= 1.2 \text{ N s}$$

b) $I = mv - mu$
$$1.2 = 0.25v - 0$$
$$v = \frac{1.2}{0.25} = 4.8 \text{ m s}^{-1}$$

Essential notes

Conservation of momentum

momentum before =
momentum after

Conservation of momentum

The principle of conservation of momentum is often applied to collisions between bodies, where the only forces acting are the forces that each body exerts on the other. This principle states that the total momentum before a collision is the same as the total momentum after a collision. This principle can also be applied in other cases such firing a gun or jumping off a skateboard, where there are no external forces acting.

The principle is derived by considering the impulse on the bodies during the collision. Figure 3.36 shows the velocities before and after a collision.

Fig. 3.36
The velocities before and after the collision.

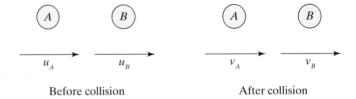

Before collision   After collision

Exam tips

Diagrams are a great help in getting momentum calculations correct.

During the collision the forces acting on the bodies are equal in magnitude and opposite in direction. This means that the impulse on A (I_A) and the impulse on B (I_B) are related by $I_A = -I_B$. Taking the masses of A and B as m_A and m_B respectively gives:

$$I_A = -I_B$$
$$m_A v_A - m_A u_A = -(m_B v_B - m_B u_B)$$
$$m_A u_A + m_B u_B = m_A v_A + m_B v_B$$

Total Momentum Before = Total Momentum After

Example

Two particles, A and B, are moving towards each other along the same straight line, when they collide. The mass of A is 4 kg and the mass of B is 6 kg. Just before the collision, the speed of A is 2.5 m s^{-1} and the speed of B is 1.2 m s^{-1}. The direction of motion of B is changed during the collision. After the collision B moves with a speed of 0.5 m s^{-1}.

a) Find the speed of A after the collision and state what happens to its direction of motion.

b) Find the magnitude of the impulse on B during the collision.

Answer

a)

Before collision   After collision

Fig. 3.37
Velocities before and after the collision.

$u_A = 2.5, m_A = 4, u_B = -1.2, m_B = 6$ and $v_B = 0.5$

$$m_A u_A + m_B u_B = m_A v_A + m_B v_B$$
$$4 \times 2.5 + 6 \times (-1.2) = 4v_A + 6 \times 0.5$$
$$2.8 = 4v_A + 3$$
$$v_A = \frac{3 - 2.8}{4} = -0.05$$

The speed is 0.05 m s^{-1}.

The direction has been reversed.

b) $I = mv - mu$
$= 6 \times 0.5 - 6 \times (-1.2)$
$= 10.2$ N s

Method notes

Figure 3.37 shows the velocities just before and just after the collision.

The values to be used in the conservation of momentum equation are first listed noting the negative initial velocity for B. The equation can then be solved. Note that the negative sign in v_A indicates that the direction of the motion has been reversed.

The impulse is calculated using $I = mv - mu$. Again note that u has a negative value.

Example

Two toy trains, which are moving along a straight horizontal track, have masses of 0.4 kg and m kg respectively. The 0.4 kg train is moving at 5 m s^{-1} when it collides with the other train which is moving at 4 m s^{-1}. Immediately after the collision the 0.4 kg train moves in the same direction at 4.2 m s^{-1} and the other train in the same direction at 4.5 m s^{-1}.

a) Find m.

b) Find the magnitude of the impulse on the 0.4 kg train.

Answer

a)

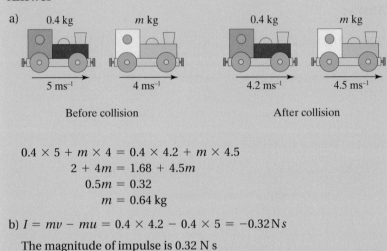

Fig.3.38
Velocities before and after the collision.

$$0.4 \times 5 + m \times 4 = 0.4 \times 4.2 + m \times 4.5$$
$$2 + 4m = 1.68 + 4.5m$$
$$0.5m = 0.32$$
$$m = 0.64 \text{ kg}$$

b) $I = mv - mu = 0.4 \times 4.2 - 0.4 \times 5 = -0.32\,\text{N}s$

The magnitude of impulse is 0.32 N s

Method notes

The velocities before and after the collision are shown in Figure 3.38.

The principle of conservation of momentum can then be applied.

In (b) the impulse is found using $I = mv - mu$.

Example

A child, of mass 25 kg, is standing on a stationary skate board, of mass 2 kg. The child jumps off the skate board. When he does this he exerts an impulse of 15 N s on the skate board. Model the child and skateboard as particles.

a) Find the velocity at which skate board moves immediately after the jump.

b) State the impulse exerted by the skate board on the child.

c) Find the velocity of the child as he jumps off the skateboard.

Answer

a) $I = mv - mu$

$15 = 2v - 2 \times 0$

$v = \dfrac{15}{2} = 7.5 \text{ m s}^{-1}$

b) The impulse on the child is -15 N s

c) $I = mv - mu$

$-15 = 25v - 25 \times 0$

$v = \dfrac{-15}{25} = -0.6 \text{ m s}^{-1}$

Method notes

The equation $I = mv - mu$ is required here. The impulse on the skate board is given. It is important to note that the impulse on the child will have the same magnitude but the opposite direction and so is -15 Ns.

Stop and think 6

A man, holding a heavy ball, stands on a trolley which has a fixed board in front of him, as shown in Figure 3.39.

Fig. 3.39
The man on the trolley at the start.

The man throws the ball at the board. When the ball hits the board it rebounds and moves backwards. Assume that the ball does not hit the man as it rebounds. Also ignore the effects of gravity.

What happens to the trolley and the ball?

Answers to Stop and think

1 a) Yes b) No c) No d) Yes

2 35 N

3 a) *The resultant force is zero because the parachutist has a constant velocity.*

b) *The resultant force is not zero. This is because although the speed is constant the velocity is not constant because its direction changes as it goes round the roundabout.*

4 *As the box moves it experiences a horizontal friction force that opposes the motion. The box also exerts a friction force on the table, which acts in the direction of motion of the box, as shown in the diagram.*

Fig. 3.40
The forces on the box and the table.

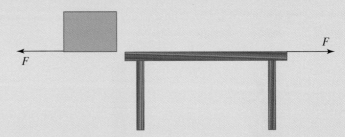

5 *The particles will both accelerate in the same direction. The string will exert the same force, say T, on each particle.*

The acceleration of the particle of mass m kg will be $\dfrac{T}{m}$ m s^{-2}.

The acceleration of the particle of mass $2m$ kg will be $\dfrac{T}{2m}$ m s^{-2}.

Thus the acceleration of the lighter particle is greater, but they both accelerate in the same direction.

6 *When the ball is first thrown, the trolley moves to the left and the ball moves to the right.*

After the ball has hit the barrier, the ball moves to the left and the trolley moves to the right.

Statics and equilibrium

The word **statics** is used to describe situations where objects remain at rest. In this section, only particles are considered and the requirement is that the forces acting on the particle are in equilibrium or have a zero resultant. A zero resultant force means that the object moves with a constant velocity, but in this section only objects at rest will be considered.

Types of force

Statics questions involve all the forces you met in the dynamics chapter. In addition there may be examples which also involve **tensions** and **thrusts**.

Tension and thrust

In the dynamics chapter we considered the tension in a string. In this section we will consider rods and springs.

If a spring is stretched then it will exert a **tension** force, directed towards the centre of the spring. However if the spring is compressed, it exerts forces away from the centre of the spring, which are called **thrusts**. These forces are illustrated in Figure 4.1, where a sphere is held in equilibrium by a spring in two different ways.

Fig. 4.1
Stretched and compressed springs.

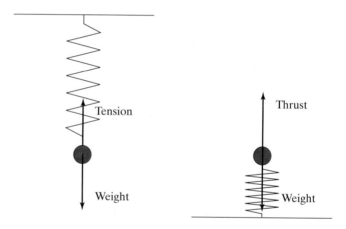

In the left hand example, the sphere is suspended using a spring. In this case the spring exerts an upward tension force on the sphere to balance the weight.

In the right hand example the spring is compressed and exerts an upward thrust to balance the weight.

It is also possible to find tensions and thrusts in rods. For example, if an object is attached to a rod and then suspended from a fixed point, the rod will exert a tension to balance the weight of the object. If a rod is used to push an object, it will exert a thrust. It is very common to find thrust forces exerted by rods. When a pole is used to support a washing line, it exerts a thrust on the line.

Using the friction inequality

When working with objects that slide, for example a child on a slide in a park, the friction force always takes the maximum possible magnitude, and so the friction can be determined using the relationship $F = \mu R$. In static situations the inequality, $F \le \mu R$, is normally used since the friction force may not take its maximum value. The exception to this is if the object is described as being on the point of slipping or sliding, in which case the maximum value for the friction can be used and problems solved using $F = \mu R$.

As an example consider a book, of mass 2 kg, that remains at rest on a table when a horizontal force of 8 N is applied to the book. The diagram in Figure 4.2 shows the forces acting on the book.

Essential notes

The friction law states that:

$F \le \mu R$ where:

F = friction

μ = coefficient of friction

R = normal reaction

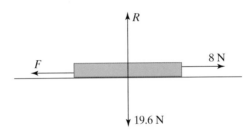

Fig. 4.2
The forces acting on the book.

Considering the vertical forces: $R = 19.6$

Considering the horizontal forces: $F = 8$

Using the friction inequality then gives:

$F \le \mu R$

$8 \le \mu \times 19.6$

$\mu \ge \dfrac{8}{19.6}$

$\mu \ge 0.408 \ldots$

$\mu \ge 0.41 \text{ (to 2 sf)}$

The value for the coefficient of friction between the book and the table cannot be found exactly, but it must be greater than or equal to 0.41.

The angle of friction

The friction law can be used to find an interesting connection between the angle of a slope or inclined plane and the coefficient of friction. Figure 4.3 shows the forces acting on a particle at rest on a rough slope inclined at an angle α to the horizontal.

Fig. 4.3
The forces acting on a particle at rest on a slope.

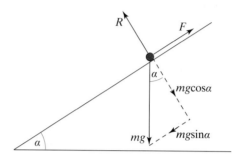

Considering the forces parallel to the slope:

$$F = mg \sin \alpha$$

Considering the forces perpendicular to the slope:

$$R = mg \cos \alpha$$

Using the friction inequality gives:

$$F \leq \mu R$$
$$mg \sin \alpha \leq \mu \times mg \cos \alpha$$
$$\mu \geq \frac{mg \sin \alpha}{mg \cos \alpha}$$
$$\mu \geq \frac{\sin \alpha}{\cos \alpha}$$
$$\mu \geq \tan \alpha$$

Essential notes

If a particle is at rest on a slope, with no forces other than the weight, friction and the normal reaction, then:

$$\mu \geq \tan \alpha$$

where α is the angle between the slope and the horizontal.

This result links the angle of the slope with the coefficient of friction. In the case where the object is on the point of slipping the coefficient of friction is equal to the tan of the angle.

Simple experiment

This experiment works well with a calculator with rubber feet. Place it on a small table or board and raise up one end of the table or board, as shown in Figure 4.4. Keep raising the table or board until the calculator is on the point of sliding. With the table or board in this position, measure the angle between the table or board and the horizontal. Use this angle to work out the coefficient of friction between the calculator and the surface using the angle of friction.

Fig. 4.4
The angle of friction experiment.

Methods for solving problems

Statics problems can be solved because the resultant of the forces acting
will be zero, which is often described as the forces being in equilibrium.
Problems are solved by using this fact and a knowledge of the properties of
the various types of force.

Using triangles of forces

If there are only three forces acting on a particle, they can be arranged to
form a triangle of forces. Once this triangle has been formed angles and
lengths, which represent forces, can be found using trigonometry. In some
cases the sine or cosine rules need to be used.

For example consider a particle that is supported by two strings, as shown
in Figure 4.5.

Fig. 4.5
A particle supported by two strings.

Three forces act on the particle, two tensions and the weight which is
vertical. These forces are shown in Figure 4.6, along with the triangle that
can be formed from them.

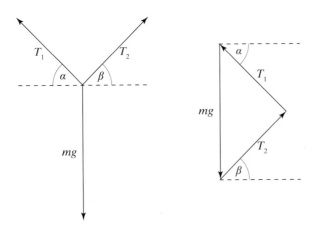

Exam tips

Learn the sine and cosine
rules in case they are needed
in an exam question.

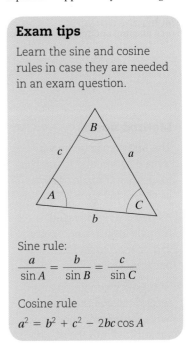

Sine rule:
$$\frac{a}{\sin A} = \frac{b}{\sin B} = \frac{c}{\sin C}$$

Cosine rule
$$a^2 = b^2 + c^2 - 2bc \cos A$$

Fig. 4.6
Creating a triangle of forces.

If the angles α and β are both known as well as the mass of the particle, then the sine rule can be used to find the tensions.

$$\frac{T_1}{\sin(90 - \beta)} = \frac{T_2}{\sin(90 - \alpha)} = \frac{mg}{\sin(\alpha + \beta)}$$

It is much easier to use and apply the sine rule in this way when the angles involved are known values. In some cases the triangle will contain a right angle and more basic trigonometry can be used.

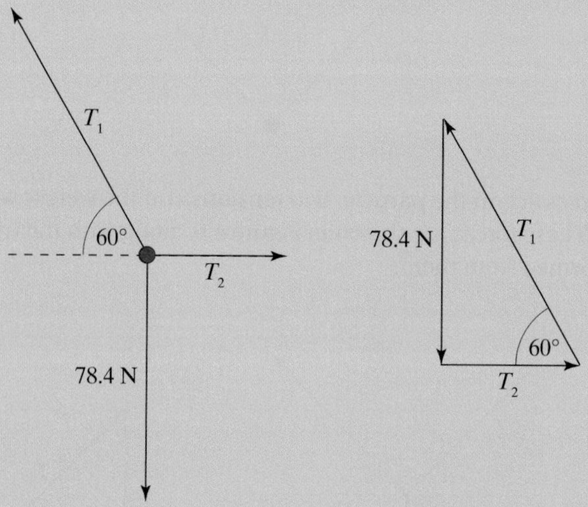

Fig. 4.7
Force diagram and triangle of forces.

Example
A particle of, mass 8 kg, is supported by two strings. One string is horizontal and the other is at an angle of 60° above the horizontal. The forces all act in the same vertical plane. Find the tension in each string.

Answer

Method notes

The first step is to draw a force diagram and to use this to create a triangle of forces as shown in Figure 4.7.

As this is a right angled triangle basic trigonometry can be used to find the two tensions.

Note that when finding the second tension the exact value for the first tension is substituted into the equation.

Also note that the final answers are given to 2 significant figures.

$T_1 \sin 60° = 78.4$

$$T_1 = \frac{78.4}{\sin 60°} = 90.5\ldots = 91 \text{ N (to 2sf)}$$

$T_1 \cos 60° = T_2$

$$T_2 = \frac{78.4}{\sin 60°} \times \cos 60° = 45.2\ldots = 45 \text{ N (to 2sf)}$$

Example

A particle, of mass 10 kg, is held at rest on a smooth inclined plane by a P N force that acts at an angle of 60° to the slope as shown in Figure 4.8. The slope is at an angle of 20° to the horizontal.

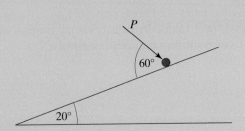

Fig. 4.8
The force of P N acting on the particle.

Find the magnitude of the reaction force acting on the particle.

Answer

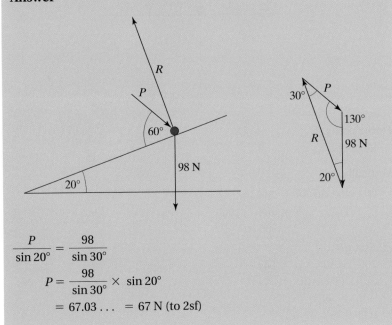

Fig. 4.9
The force diagram and triangle of forces.

$$\frac{P}{\sin 20°} = \frac{98}{\sin 30°}$$

$$P = \frac{98}{\sin 30°} \times \sin 20°$$

$$= 67.03 \ldots = 67 \text{ N (to 2sf)}$$

Method notes

Figure 4.9 shows the forces acting on the particle and the triangle of forces.

Note that the angle between the force P and the weight has been found $(60 + 70 = 130)$.

The angle between P and R has been found.

$(180 - 130 - 20 = 30°)$

The sine rule has been used to find P.

Using components of forces

If a problem involves more than three forces, then the triangle of forces method cannot be used. The equivalent method would be to use a quadrilateral of forces. While this method would eventually lead to a solution, it would be cumbersome. A better approach is to use the components of the forces.

The components should be taken in two perpendicular directions, normally horizontal and vertical or parallel and perpendicular to a slope. The components in both directions must be in equilibrium. Often these equations will lead to a pair of simultaneous equations.

Essential notes

For the forces on the particle to be in equilibrium, the components in both directions (for example horizontal and vertical) must be in equilibrium.

Example

A crate, of mass 80 kg, is at rest on a rough horizontal surface. A force of magnitude 200 N acts on the crate at an angle of 30° above the horizontal. The coefficient of friction between the surface and the crate is μ.

Model the crate as a particle.

Show that, correct to two significant figures, $\mu \geq 0.25$.

Answer

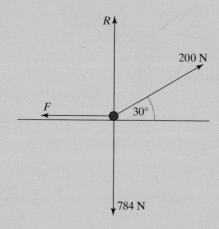

Fig. 4.10
Forces on the crate.

Method notes

Figure 4.10 shows the forces acting on the crate.

The 200 N force must be split into horizontal and vertical components, which can then be used to form to equations, to find R and F.

The inequality $F \leq \mu R$ can then be used to show the required result.

Horizontally:

$F = 200 \cos 30°$

Vertically:

$R + 200 \sin 30° = 784$

$R = 784 - 100$

$= 684$ N

$F \leq \mu R$

$200 \cos 30° \leq \mu \times 684$

$\mu \geq \dfrac{200 \cos 30°}{684}$

$\mu \geq 0.253 \ldots$

$\mu \geq 0.25$ (to 2sf)

Example

A particle, of mass 5 kg, is held at rest on a rough inclined plane by a horizontal force of P N as shown in Figure 4.11. The plane is at an angle of 40° to the horizontal and the coefficient of friction between the particle and the plane is 0.3.

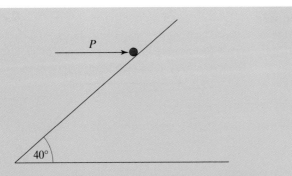

Fig. 4.11
The horizontal force P.

Find the largest value of P for which the particle remains at rest.

Answer

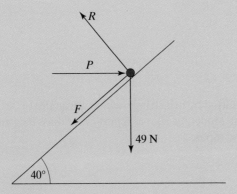

Fig. 4.12
The forces on the particle when it is on the point of sliding up the plane.

Parallel to the plane

$$P \cos 40° = F + 49 \sin 40°$$

$$F = P \cos 40° - 49 \sin 40°$$

Perpendicular to the plane

$$R = P \sin 40° + 49 \cos 40°$$

Since the particle is on the point of sliding then:

$$F = \mu R$$
$$P \cos 40° - 49 \sin 40° = 0.3 \times (P \sin 40° + 49 \cos 40°)$$
$$P \cos 40° - 0.3P \sin 40° = 0.3 \times 49 \cos 40° + 49 \sin 40°$$
$$P(\cos 40° - 0.3 \sin 40°) = 14.7 \cos 40° + 49 \sin 40°$$
$$P = \frac{14.7 \cos 40° + 49 \sin 40°}{\cos 40° - 0.3 \sin 40°}$$
$$= 74.5 \ldots = 75 \text{ N (to 2sf)}$$

> **Method notes**
>
> When P takes its maximum value, the particle will be at rest, but on the point of moving up the plane. This means that the friction must act down the plane and that the friction must take its maximum value of μR. The forces are shown in Figure 4.12.
>
> Considering the components parallel and perpendicular to the plane gives expressions for R and F in terms of P.
>
> These can then be used with $F = \mu R$ to find P.
>
> Note that the first key step in solving the equation is to bring all the terms with a P to the left hand side of the equation.

Lami's Theorem

Lami's Theorem is a result that can be used when solving problems in statics. It applies when three forces act on a particle in equilibrium. The forces must all lie in a single plane. Figure 4.13 shows the forces and the angles used in **Lami's Theorem**.

Fig. 4.13
The arrangement of forces and angles for Lami's Theorem.

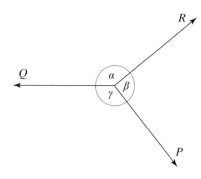

Lami's Theorem states that:

$$\frac{P}{\sin \alpha} = \frac{Q}{\sin \beta} = \frac{R}{\sin \gamma}$$

Fig. 4.14
The forces and angles for use with Lami's Theorem.

Method notes

The first diagram in Figure 4.14 shows the forces and angles give in the question.

The angle between the two tensions is:

$180 - 30 - 40 = 110°$

The angle between the weight and T_1 is $90 + 30 = 120°$

The angle between the weight and T_2 is $90 + 40 = 130°$

Lami's Theorem can then be applied with these angles and the forces.

Example
A sign is supported by two strings, which are angles of 30° and 40° to the horizontal. Model the sign as a particle of mass 8 kg and find the tensions in the ropes.

Answer

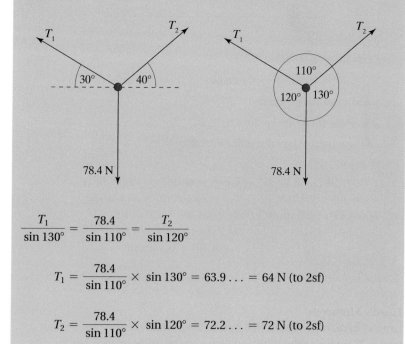

$$\frac{T_1}{\sin 130°} = \frac{78.4}{\sin 110°} = \frac{T_2}{\sin 120°}$$

$$T_1 = \frac{78.4}{\sin 110°} \times \sin 130° = 63.9 \ldots = 64 \text{ N (to 2sf)}$$

$$T_2 = \frac{78.4}{\sin 110°} \times \sin 120° = 72.2 \ldots = 72 \text{ N (to 2sf)}$$

Example

The diagram in Figure 4.15 shows three forces that are in equilibrium. The 40 N force is horizontal and the other two have magnitude 25 N and P N and act at angles of α and 20° to the horizontal, as shown.

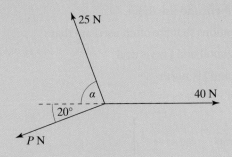

Fig. 4.15
The three forces.

a) Find α.

b) Find P.

Answer

a)
$$\frac{25}{\sin 160°} = \frac{40}{\sin (\alpha + 20)}$$
$$\sin (\alpha + 20) = \frac{40 \sin 160°}{25}$$
$$\alpha + 20 = 33.2°$$
$$\alpha = 33.2 - 20 = 13.2° = 13° \text{ (to 2sf)}$$

b)
$$\frac{25}{\sin 160°} = \frac{P}{\sin (180 - 13.2)}$$
$$P = \frac{25}{\sin 160°} \times \sin (180 - 13.2)$$
$$P = 16.6\ldots = 17 \text{ N (to 2sf)}$$

Method notes

First apply Lami's Theorem with the 25 N and 40 N forces. Note that the angles required are:

$20 + \alpha$ between the P N and 25 N forces and

$180 - 20 = 160°$ between the P N and 40 N forces.

Lami's Theorem can be applied to find α in (a) and P in (b), where the value of α found in (a) needs to be used.

Stop and think

A particle, of mass m kg, is at rest on a rough slope, inclined at an angle to the horizontal. The magnitude of the normal reaction force is R N. Which of the following statements is true?

A: $R = mg$

B: $R > mg$

C: $R < mg$

Stop and think answers

The statement C is correct. Resolving perpendicular to the slope gives $R = mg \cos \theta$, where θ is the angle between the slope and the horizontal. As $\cos \theta < 1$ for acute slope angles θ then $R < mg$.

Unit vectors in mechanics

Vectors are often used in mechanics to express forces, displacements, velocities and accelerations. Two perpendicular **unit vectors i** and **j** are normally used in mechanics, as shown in Figure 5.1. The important principle is that the unit vectors are perpendicular which makes it easy to work with them and use them.

Two conventions that are often used are that:

- **i** is horizontal and **j** is vertical
- **i** is east and **j** is north.

Fig. 5.1
Unit vectors.

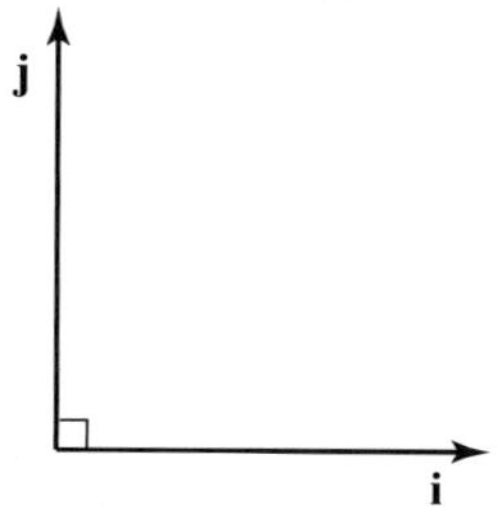

Essential notes

Unit vectors have magnitude 1. For example, this could be 1 metre when working with displacements or 1 newton when working with forces.

Position vectors and displacement

The grid in Figure 5.2 shows the unit vectors **i** and **j** which are directed east and north respectively. The point marked O is the origin. A boy called Alan is at the point A and his friend Bill is at the point B.

Fig. 5.2
The positions of A and B.

Essential notes

The position vector uses the unit vectors to describe where an object is located. Position vectors of moving object are often expressed in terms of time t.

The **position vectors** of A and B can be described with respect to the origin O.

Position Vector of $A = 4\mathbf{i} + \mathbf{j}$

Position Vector of $B = 9\mathbf{i} + 5\mathbf{j}$

These position vectors describe where an object is with reference to an origin. The term displacement is used to describe a change of position. Consider the displacements described in the examples below with reference to Figure 5.2.

If a particle moves from A to B its displacement would be $5\mathbf{i} + 4\mathbf{j}$. This can be seen directly from the diagram or calculated using:

$\mathbf{r}_B - \mathbf{r}_A = 9\mathbf{i} + 5\mathbf{j} - (4\mathbf{i} + \mathbf{j}) = 5\mathbf{i} + 4\mathbf{j}.$

- If a particle moves from B to A its displacement would be $-5\mathbf{i} - 4\mathbf{j}$.
- If a particle moves from A to O its displacement would be $-4\mathbf{i} - \mathbf{j}$.

If a particle, or other object, is moving with a certain velocity for a specified period of time, then the change in position, or displacement, can be calculated easily.

New position vectors can also be found by adding the displacement to the original position vector, provided that the velocity and the time for which the object are moving are both known.

> **Example**
> A boat has position vector $(40\mathbf{i} + 250\mathbf{j})$ m with respect to an origin at a lighthouse. It moves for 240 seconds with a constant velocity of $(5\mathbf{i} - 2\mathbf{j})$ ms^{-1}.
>
> a) Find the displacement of the boat during the 240 seconds.
>
> b) Find the distance travelled during the 240 seconds.
>
> c) Find the distance of the boat from the light house at the end of the 240 seconds.
>
> **Answer**
> a) Displacement $= (5\mathbf{i} - 2\mathbf{j}) \times 240 = 1200\mathbf{i} - 480\mathbf{j}$
>
> b)
>
>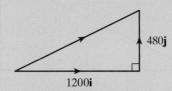
>
> $$\text{Distance travelled} = \sqrt{1200^2 + 480^2}$$
> $$= 1292 \text{ m (to the nearest m)}$$
>
> c) After 240 seconds:
> $$\mathbf{r} = (5\mathbf{i} - 2\mathbf{j}) \times 240 + (40\mathbf{i} + 250\mathbf{j})$$
> $$= 1200\mathbf{i} - 480\mathbf{j} + 40\mathbf{i} + 250\mathbf{j}$$
> $$= 1240\mathbf{i} - 230\mathbf{j}$$
>
>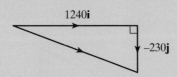
>
> $$\text{Distance from lighthouse} = \sqrt{1240^2 + 230^2}$$
> $$= 1261 \text{ m (to the nearest m)}$$

Essential notes

For motion with a constant velocity:

Displacement $= \mathbf{v}\,t$

where

$\mathbf{v} =$ velocity and $t =$ time

Essential notes

For motion with a constant velocity:

$\mathbf{r} = \mathbf{v}\,t + \mathbf{r}_0$

where:

$\mathbf{v} =$ velocity, $t =$ time and

$\mathbf{r}_0 =$ initial position

Fig. 5.3
The distance is given by the hypotenuse of this triangle

Method notes

In (a), the displacement is found by multiplying the velocity by the time.

In (b), the distance travelled is the magnitude of the displacement vector.

In (c), the position vector relative to the lighthouse must be found first using the formula

$\mathbf{r} = \mathbf{v}\,t + \mathbf{r}_0$

The magnitude of the position vector gives the distance from the lighthouse.

Fig. 5.4
The distance is given by the hypotenuse of this triangle

Method notes

The displacement is the difference between the two position vectors.

Once the displacement has been found the velocity can be found using the formula:

displacement = $\mathbf{v}t$

Example

A particle moves with a constant velocity for 10 seconds. At the beginning of the 10 second period it has position vector $(56\mathbf{i} + 72\mathbf{j})$ m and at the end it has position vector $(102\mathbf{i} + 134\mathbf{j})$ m.

Find the velocity of the particle.

Answer

Displacement $= (102\mathbf{i} + 134\mathbf{j}) - (56\mathbf{i} + 72\mathbf{j})$
$$= 46\mathbf{i} + 62\mathbf{j} \text{ m}$$

$$46\mathbf{i} + 62\mathbf{j} = 10\mathbf{v}$$
$$\mathbf{v} = \frac{46\mathbf{i} + 62\mathbf{j}}{10} = 4.6\mathbf{i} + 6.2\mathbf{j}$$

Essential notes

The magnitude of the vector $a\mathbf{i} + b\mathbf{j}$ is given by $\sqrt{a^2 + b^2}$.

As any number squared is always positive, there is no need to take account of any negative signs when calculating the magnitude of a vector.

Fig. 5.5
Finding the magnitude of the vector $a\mathbf{i} + b\mathbf{j}$.

Magnitude and direction of a vector

The **magnitude** of a vector gives the size of the vector. In some instances this will represent a length, but for a velocity vector the magnitude would give the speed. For a force vector the magnitude of the vector would give the size of the force. Consider the vector $a\mathbf{i} + b\mathbf{j}$. The magnitude of the vector is given by the hypotenuse of the right angled triangle shown in Figure 5.5 and can be found using Pythagoras' Theorem.

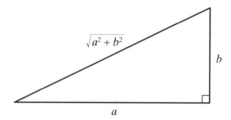

It is also useful to be able to find the direction of a vector. This is usually given as an angle with reference to one of the unit vectors or maybe as a bearing if the position of an object is being considered.

Essential notes

When finding the direction of a vector, use tan to find an angle and then think carefully about how to specify the direction.

The direction can be found using trigonometry, usually done using the relationship $\tan \theta = \dfrac{\text{Opposite}}{\text{Adjacent}}$. This is illustrated in Figure 5.6, where the angle θ is marked on the diagram. In this case $\tan \theta = \dfrac{b}{a}$ and the angle θ will be measured anti-clockwise from the direction unit vector $\mathbf{i}$.

Fig. 5.6
Finding the direction of the vector $a\mathbf{i} + b\mathbf{j}$.

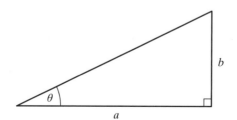

Example

a) Find the magnitude of the vector $4\mathbf{i} + 7\mathbf{j}$.

b) Find the angle between the vector $4\mathbf{i} + 7\mathbf{j}$ and the unit vector $\mathbf{i}$.

Answer

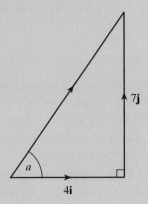

Fig. 5.7
The vector $4\mathbf{i} + 7\mathbf{j}$.

Method notes

The vector is illustrated in Figure 5.7.

For (a) the magnitude is calculated using Pythagoras' Theorem.

For (b), the angle is found using $\tan\theta = \dfrac{Opposite}{Adjacent}$.

a) Magnitude of $4\mathbf{i} + 7\mathbf{j} = \sqrt{4^2 + 7^2} = 8.1$ N (to 2 sf)

b) $\tan\alpha = \dfrac{7}{4}$

$\alpha = \tan^{-1}\left(\dfrac{7}{4}\right) = 60.3°$ (to 1 dp)

Exam tips

Ignore the negative signs when calculating magnitudes as mistakes can be made by not entering negative signs correctly into calculators.

Example

a) Find the magnitude of the vector $-3\mathbf{i} + 2\mathbf{j}$.

b) Find the angle between the vector $-3\mathbf{i} + 2\mathbf{j}$ and the unit vector $\mathbf{i}$.

Answer

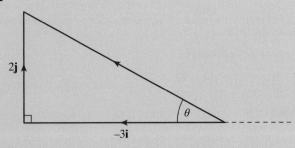

Fig. 5.8
The vector $-3\mathbf{i} + 2\mathbf{j}$.

Method notes

The vector can be drawn as shown in Figure 5.8.

The magnitude of the vector can be found using Pythagoras' Theorem.

To find the angle between the vector and the unit vector, first find the acute angle between the vector and the $-\mathbf{i}$ direction. Then subtract this from 180 to find the angle between the $\mathbf{i}$ direction.

a) Magnitude $\sqrt{3^2 + 2^2} = \sqrt{13} = 3.6$ (to 2 sf)

b) $\tan\theta = \dfrac{2}{3}$

$\theta = \tan^{-1}\left(\dfrac{2}{3}\right) = 33.7°$ (to 1 dp)

Angle between the vector and unit vector $\mathbf{i} = 180 - 33.7 = 146.3°$

Solving problems with position vectors

Vectors can be used to solve a range of problems. These will often involve the motion of two objects and might require finding out if two objects collide or the conditions under which they would collide. This often involves setting up expressions for the position vectors of the two objects in terms of time and then finding out if they both have the same position vector at the same time.

Example

Two particles A and B are moving on the same horizontal surface. Particle A starts at the point with position vector $8\mathbf{i} + 12\mathbf{j}$ m and moves with a constant velocity of $2\mathbf{i} - 5\mathbf{j}$ m s^{-1}.

Particle B starts at the point with position vector $-3\mathbf{i} + 23\mathbf{j}$ m and moves with a constant velocity of $3\mathbf{i} - 6\mathbf{j}$ m s^{-1}.

Find the position vector of the point where the two particles collide.

Answer

Position vector of A:

$$\begin{aligned} \mathbf{r}_A &= (8\mathbf{i} + 12\mathbf{j}) + (2\mathbf{i} - 5\mathbf{j})t \\ &= 8\mathbf{i} + 12\mathbf{j} + 2t\mathbf{i} - 5t\mathbf{j} \\ &= (8 + 2t)\mathbf{i} + (12 - 5t)\mathbf{j} \end{aligned}$$

Position vector of B:

$$\begin{aligned} \mathbf{r}_B &= (-3\mathbf{i} + 23\mathbf{j}) + (3\mathbf{i} - 6\mathbf{j})t \\ &= -3\mathbf{i} + 23\mathbf{j} + 3t\mathbf{i} - 6t\mathbf{j} \\ &= (-3 + 3t)\mathbf{i} + (23 - 6t)\mathbf{j} \end{aligned}$$

For a collision:

$$8 + 2t = -3 + 3t \quad \text{and} \quad 12 - 5t = 23 - 6t$$
$$t = 11 \qquad \text{and} \qquad t = 11$$

Collide when $t = 11$.

Position vector for collision:

$$\begin{aligned} \mathbf{r}_B &= (-3\mathbf{i} + 23\mathbf{j}) + (3\mathbf{i} - 6\mathbf{j}) \times 11 \\ &= -3\mathbf{i} + 23\mathbf{j} + 33\mathbf{i} - 66\mathbf{j} \\ &= 30\mathbf{i} - 43\mathbf{j} \end{aligned}$$

Method notes

First the position vectors must both be found in terms of time t.

If the particles collide, then both the $\mathbf{i}$ components and both the $\mathbf{j}$ components must be equal at the same time. This allows two equations to be formed. In this case both give the same two values for t, so the two particles collide when $t = 11$.

Finally calculate the position vector at this time using either of the position vectors. (Both should give the same answer.)

Example

A patrol boat wants to intercept a boat that it thinks is involved in smuggling. At noon the smugglers' boat has position vector $20\mathbf{i} + 31\mathbf{j}$ km and moves with a constant velocity of $3\mathbf{i} + 2\mathbf{j}$ km h^{-1}. At noon the patrol boat has position vector $-7\mathbf{i} + 4\mathbf{j}$ km and starts to move with a constant velocity of $k\mathbf{i} + 8\mathbf{j}$ km h^{-1}. The unit vectors $\mathbf{i}$ and $\mathbf{j}$ are directed east and north respectively.

a) Find the value of k for which the patrol boat will intercept the smugglers.

b) State the time when the patrol boat intercepts the smugglers.

c) Find the distance travelled by the patrol boat as it travels to intercept the smugglers.

d) Find the bearing on which the patrol boat is travelling.

Answer

a) Position vector of the smugglers:

$$\mathbf{r}_S = (20\mathbf{i} + 31\mathbf{j}) + (3\mathbf{i} + 2\mathbf{j})t$$
$$= 20\mathbf{i} + 31\mathbf{j} + 3t\mathbf{i} + 2t\mathbf{j}$$
$$= (20 + 3t)\mathbf{i} + (31 + 2t)\mathbf{j}$$

Position vector of the patrol boat:

$$\mathbf{r}_P = (-7\mathbf{i} + 4\mathbf{j}) + (k\mathbf{i} + 8\mathbf{j})t$$
$$= -7\mathbf{i} + 4\mathbf{j} + kt\mathbf{i} + 8t\mathbf{j}$$
$$= (-7 + kt)\mathbf{i} + (4 + 8t)\mathbf{j}$$

For interception:

$$20 + 3t = -7 + kt \quad \text{and} \quad 31 + 2t = 4 + 8t$$

From the second equation:

$$31 + 2t = 4 + 8t$$
$$27 = 6t$$
$$t = \frac{27}{6} = 4.5$$

Using the first equation:

$$20 + 3t = -7 + kt$$
$$20 + 3 \times 4.5 = -7 + 4.5k$$
$$33.5 = -7 + 4.5k$$
$$40.5 = 4.5k$$
$$k = \frac{40.5}{4.5} = 9$$

b) Patrol boat intercepts the smugglers 4.5 hours after noon, that is at 1630 or 4.30 pm.

Method notes

First find the position vectors of both boats in terms of t and k.

For the patrol boat to intercept the smugglers, both must have equal $\mathbf{i}$ components and equal $\mathbf{j}$ components. This allows two equations to be formed. Solving the one that only contains t shows that the interception must take place after 4.5 hours.

This value can then be substituted into the other equation to find the value of k.

In (b), note that an actual time is expected as the answer.

Fig. 5.9
Finding the bearing.

c) Displacement of patrol boat = $(9\mathbf{i} + 8\mathbf{j}) \times 4.5$

$$= 40.5\mathbf{i} + 36\mathbf{j}$$

Distance = $\sqrt{40.5^2 + 36^2}$

$$= 54 \text{ km (correct to the nearest km)}$$

d)

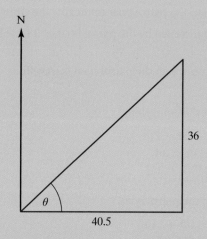

The angle θ is given by:

$$\theta = \tan^{-1}\left(\frac{36}{40.5}\right) = 41.6° \text{ (to 1 dp)}$$

Bearing $= 90 - 41.6 = 48.4°$

Velocity vectors

Velocity vectors are often expressed in the form $a\mathbf{i} + b\mathbf{j}$, but in some cases can be expressed in different ways. One that you may encounter is when a velocity is given in terms of a speed v and a direction **d**. For example you might be told that a helicopter has a speed of 60 m s^{-1} in the direction $3\mathbf{i} + 4\mathbf{j}$. To solve problems like this a unit vector in the direction of **d** is needed. This is found by dividing the vector by its magnitude. In this case the unit vector would be:

$$\frac{3\mathbf{i} + 4\mathbf{j}}{\sqrt{3^2 + 4^2}} = \frac{3\mathbf{i} + 4\mathbf{j}}{5} = 0.6\mathbf{i} + 0.8\mathbf{j}$$

Example

A particle moves with speed $5\sqrt{10}$ m s^{-1} in the direction $3\mathbf{i} + 9\mathbf{j}$. Find the velocity of the particle in the form $a\mathbf{i} + b\mathbf{j}$.

Answer

$$\mathbf{v} = 5\sqrt{10} \times \frac{3\mathbf{i} + 9\mathbf{j}}{\sqrt{3^2 + 9^2}}$$

$$= 5\sqrt{10} \times \frac{3\mathbf{i} + 9\mathbf{j}}{\sqrt{90}}$$

$$= 5\sqrt{10} \times \frac{3\mathbf{i} + 9\mathbf{j}}{3\sqrt{10}}$$

$$= 5\mathbf{i} + 15\mathbf{j}$$

Method notes

The velocity is obtained by multiplying a unit vector in the direction of $3\mathbf{i} + 9\mathbf{j}$ by the speed.

Method notes

The velocity is obtained by multiplying a unit vector in the direction of $4\mathbf{i} - 8\mathbf{j}$ by the speed.

In this case note that

$80 = 16 \times 5$, so that

$$\sqrt{80} = \sqrt{16} \times \sqrt{5} = 4\sqrt{5}$$

The displacement is calculated using displacement $= \mathbf{v}t$.

Example

A particle moves with speed $6\sqrt{5}$ m s^{-1} in the direction $4\mathbf{i} - 8\mathbf{j}$ for 20 seconds. Find the displacement of the particle after the 20 seconds.

Answer

$$\mathbf{v} = 6\sqrt{5} \times \frac{4\mathbf{i} - 8\mathbf{j}}{\sqrt{4^2 + 8^2}}$$

$$= 6\sqrt{5} \times \frac{4\mathbf{i} - 8\mathbf{j}}{\sqrt{80}}$$

$$= 6\sqrt{5} \times \frac{4\mathbf{i} - 8\mathbf{j}}{4\sqrt{5}}$$

$$= 6\mathbf{i} - 12\mathbf{j}$$

Displacement $= (6\mathbf{i} - 12\mathbf{j}) \times 20$
$\qquad\qquad\quad = 120\mathbf{i} - 240\mathbf{j}$

Stop and think 1

The unit vectors $\mathbf{i}$ and $\mathbf{j}$ are directed east and north respectively. In what direction is a boat travelling if it has velocity:

a) $-5\mathbf{i}$

b) $6\mathbf{j}$

c) $8\mathbf{i} + 8\mathbf{j}$

d) $-2\mathbf{i} + 2\mathbf{j}$

e) $-4\mathbf{j}$

f) $3\mathbf{i} - 3\mathbf{j}$

Acceleration and velocity

It is possible link the velocity and acceleration vectors using the constant acceleration equation $\mathbf{v} = \mathbf{u} + \mathbf{a}t$. This is the vector equivalent to the equation $v = u + at$ that was used for motion on a straight line.

Essential notes

The equation $\mathbf{v} = \mathbf{u} + \mathbf{a}t$ can be used where the acceleration is constant and where:

$\mathbf{a}$ = acceleration

$\mathbf{u}$ = initial velocity

$\mathbf{v}$ = final velocity

t = time.

Example
A particle accelerates for 20 seconds. During this time the acceleration is constant and the velocity changes from $(5\mathbf{i} - 2\mathbf{j})$ m s^{-1} to $(20\mathbf{i} + 12\mathbf{j})$ m s^{-1}. Find the acceleration of the particle.

Answer
$\mathbf{v} = \mathbf{u} + \mathbf{a}t$ with $\mathbf{v} = 20\mathbf{i} + 12\mathbf{j}$, $\mathbf{u} = 5\mathbf{i} - 2\mathbf{j}$ and $t = 20$:

$$20\mathbf{i} + 12\mathbf{j} = 5\mathbf{i} - 2\mathbf{j} + 20\mathbf{a}$$
$$15\mathbf{i} + 14\mathbf{j} = 20\mathbf{a}$$
$$\mathbf{a} = \frac{15\mathbf{i} + 14\mathbf{j}}{20}$$
$$= (0.75\mathbf{i} + 0.7\mathbf{j}) \text{ m s}^{-2}$$

Method notes

The values can be substituted into the equation $\mathbf{v} = \mathbf{u} + \mathbf{a}t$ and rearranged to find $\mathbf{a}$.

Method notes

The values given can be substituted into the equation $\mathbf{v} = \mathbf{u} + \mathbf{a}t$ to find $\mathbf{v}$.

Once the velocity has been found, the speed is given by the magnitude of the velocity.

Example
A particle has velocity $(-8\mathbf{i} - 9\mathbf{j})$ m s^{-1} when it begins to accelerate. It accelerates at $(0.4\mathbf{i} + 0.6\mathbf{j})$ m s^{-2} for 40 seconds. Find the speed of the particle at the end of the 40 seconds.

Answer

$\mathbf{v} = \mathbf{u} + \mathbf{a}t$ with $\mathbf{u} = -8\mathbf{i} - 9\mathbf{j}$, $\mathbf{a} = 0.4\mathbf{i} + 0.6\mathbf{j}$ and $t = 40$:
$$\mathbf{v} = -8\mathbf{i} - 9\mathbf{j} + 40(0.4\mathbf{i} + 0.6\mathbf{j})$$
$$= -8\mathbf{i} - 9\mathbf{j} + 16\mathbf{i} + 24\mathbf{j}$$
$$= (8\mathbf{i} + 15\mathbf{j}) \text{ m s}^{-1}$$

$$\text{Speed} = \sqrt{8^2 + 15^2}$$
$$= 17 \text{ m s}^{-1}$$

Exam tips

Look carefully at questions to see if they are asking for the velocity or the speed.

Forces and vectors

Forces that act in a plane can be represented as vectors using the perpendicular unit vectors $\mathbf{i}$ and $\mathbf{j}$. It is very easy to add these vectors together to find resultants.

Forces in equilibrium
If the forces are in equilibrium, then the resultant of the force will be zero. This allows problems that are expressed in vector form to be solved very easily.

Essential notes

If the forces a on a body are in equilibrium, then the resultant is $0\mathbf{i} + 0\mathbf{j}$.

Method notes

To show that the forces are in equilibrium, it is enough to show that the resultant force is zero.

Example

The forces $\mathbf{P} = 6\mathbf{i} + 7\mathbf{j}$, $\mathbf{Q} = 2\mathbf{i} - 3\mathbf{j}$ and $\mathbf{R} = -8\mathbf{i} - 4\mathbf{j}$ act on a particle. Show that these forces are in equilibrium.

Answer

$$\begin{aligned}\mathbf{P} + \mathbf{Q} + \mathbf{R} &= 6\mathbf{i} + 7\mathbf{j} + 2\mathbf{i} - 3\mathbf{j} - 8\mathbf{i} - 4\mathbf{j} \\ &= (6 + 2 - 8)\mathbf{i} + (7 - 3 - 4)\mathbf{j} \\ &= 0\mathbf{i} + 0\mathbf{j}\end{aligned}$$

As the resultant is zero the forces must be in equilibrium.

Example

The forces $\mathbf{F}_1 = 3\mathbf{i} - 7\mathbf{j}$, $\mathbf{F}_2 = 8\mathbf{i} + 2\mathbf{j}$ and $\mathbf{F}_3$ are in equilibrium.

a) Find $\mathbf{F}_3$.

b) Find the magnitude of $\mathbf{F}_3$.

c) Find the angle between $\mathbf{F}_3$ and the unit vector $\mathbf{i}$.

Answer

a)
$$\begin{aligned}\mathbf{F}_1 + \mathbf{F}_2 + \mathbf{F}_3 &= 0\mathbf{i} + 0\mathbf{j} \\ 3\mathbf{i} - 7\mathbf{j} + 8\mathbf{i} + 2\mathbf{j} + \mathbf{F}_3 &= 0\mathbf{i} + 0\mathbf{j} \\ 11\mathbf{i} - 5\mathbf{j} + \mathbf{F}_3 &= 0\mathbf{i} + 0\mathbf{j} \\ \mathbf{F}_3 &= -11\mathbf{i} + 5\mathbf{j}\end{aligned}$$

b)

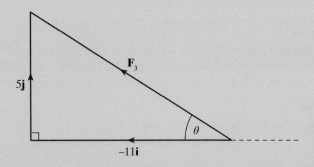

Magnitude of $\mathbf{F}_3 = \sqrt{11^2 + 5^2} = \sqrt{146} = 12.1$ (to 3 sf)

c) $\tan \theta = \dfrac{5}{11}$

$\theta = \tan^{-1}\left(\dfrac{5}{11}\right) = 24.4°$ (to 1 dp)

Angle between $\mathbf{F}_3$ and the unit vector $\mathbf{i} = 180 - 24.4 = 155.6°$

Exam tip

In 'show that' questions like this make sure that you do show all of the working.

Method notes

The sum of the three forces must be zero as they are in equilibrium.

Figure 5.10 shows the forces $\mathbf{F}_3$. From this diagram the magnitude can be found using Pythagoras' Theorem and the angle can be found using trigonometry.

Fig. 5.10
The force $\mathbf{F}_3$.

Essential notes

Newton's Second Law can be expressed in vector form as $\mathbf{F} = m\mathbf{a}$, where:

$\mathbf{F}$ = resultant force

m = mass

$\mathbf{a}$ = acceleration.

Method notes

The first step is to find the resultant force, which in this case is expressed as a vector.

Newton's Second Law can then be applied to find the acceleration.

Method notes

The resultant force can be found in terms of $\mathbf{R}$. Newton's Second Law can then be applied using the known acceleration to find $\mathbf{R}$.

Finally Pythagoras' theorem can be used to find the magnitude of the force.

Vectors and Newton's Second Law

Vectors can be used in conjunction with Newton's Second law where the forces and acceleration are expressed as vectors.

Example

Find the acceleration when the following forces act on a particle of mass 5 kg:

$$4\mathbf{i} + 5\mathbf{j}$$
$$6\mathbf{i} - 2\mathbf{j}$$
$$-12\mathbf{i} + 3\mathbf{j}$$
$$4\mathbf{i} + 14\mathbf{j}$$

Answer

$$\begin{aligned}
\text{Resultant force} &= 4\mathbf{i} + 5\mathbf{j} + 6\mathbf{i} - 2\mathbf{j} - 12\mathbf{i} + 3\mathbf{j} + 4\mathbf{i} + 14\mathbf{j} \\
&= (4 + 6 - 12 + 4)\mathbf{i} + (5 - 2 + 3 + 14)\mathbf{j} \\
&= 2\mathbf{i} + 20\mathbf{j}
\end{aligned}$$

Applying Newton's Second Law:

$$\mathbf{F} = m\mathbf{a}$$
$$2\mathbf{i} + 20\mathbf{j} = 5\mathbf{a}$$
$$\mathbf{a} = \frac{2\mathbf{i} + 20\mathbf{j}}{5} = 0.4\mathbf{i} + 4\mathbf{j} \text{ m s}^{-2}$$

Example

Three forces, $\mathbf{P}$, $\mathbf{Q}$ and $\mathbf{R}$, act on a particle of mass 7 kg to produce an acceleration of $0.8\mathbf{i} + 0.6\mathbf{j}$ m s^{-2}. If $\mathbf{P} = 6\mathbf{i} + 10\mathbf{j}$ N and $\mathbf{Q} = 4\mathbf{i} + 8\mathbf{j}$ N, find the magnitude of $\mathbf{R}$, giving your answer correct to the nearest newton.

Answer

$$\begin{aligned}
\text{Resultant Force} &= 6\mathbf{i} + 10\mathbf{j} + 4\mathbf{i} + 8\mathbf{j} + \mathbf{R} \\
&= 10\mathbf{i} + 18\mathbf{j} + \mathbf{R}
\end{aligned}$$

Applying Newton's Second Law:

$$\mathbf{F} = m\mathbf{a}$$
$$10\mathbf{i} + 18\mathbf{j} + \mathbf{R} = 7(0.8\mathbf{i} + 0.6\mathbf{j})$$
$$10\mathbf{i} + 18\mathbf{j} + \mathbf{R} = 5.6\mathbf{i} + 4.2\mathbf{j}$$
$$\mathbf{R} = 5.6\mathbf{i} + 4.2\mathbf{j} - 10\mathbf{i} - 18\mathbf{j}$$
$$= -4.4\mathbf{i} - 13.8\mathbf{j}$$

Magnitude of $\mathbf{R} = \sqrt{4.4^2 + 13.8^2} = 14$ N (to the nearest N)

Example

A resultant force of $(22\mathbf{i} + 20\mathbf{j})$ N acts on a particle of mass 4 kg for 6 seconds. The particle was moving with velocity $(3\mathbf{i} - 3\mathbf{j})$ m s^{-1} when the force began to act. Find the speed of the particle at the end of the 6 seconds.

Answer

$$\mathbf{F} = m\mathbf{a}$$

$$22\mathbf{i} + 20\mathbf{j} = 4\mathbf{a}$$

$$\mathbf{a} = \frac{22\mathbf{i} + 20\mathbf{j}}{4} = 5.5\mathbf{i} + 5\mathbf{j} \text{ m s}^{-2}$$

$$\mathbf{v} = \mathbf{u} + \mathbf{a}t$$
$$= 3\mathbf{i} - 3\mathbf{j} + 6(5.5\mathbf{i} + 5\mathbf{j})$$
$$= 3\mathbf{i} - 3\mathbf{j} + 33\mathbf{i} + 30\mathbf{j}$$
$$= 36\mathbf{i} + 27\mathbf{j}$$

$$\text{Speed} = \sqrt{36^2 + 27^2}$$
$$= 45 \text{ m s}^{-1}$$

Method notes

The first stage of the solution is to apply Newton's Second Law to obtain the acceleration vector.

The equation $\mathbf{v} = \mathbf{u} + \mathbf{a}t$ can then be used to find the final velocity.

The speed, which is the magnitude of the velocity, can then be found.

Stop and think 2

The resultant force, **F**, on a particle is given by:

$$\mathbf{F} = (4 - p)\mathbf{i} + (q - 2)\mathbf{j}$$

where the unit vectors **i** and **j** are horizontal and vertical respectively.

What can be deduced about the values of p and q if:

a) The resultant force is zero.

b) The resultant force is vertical.

c) The resultant force is horizontal.

d) The resultant force acts in a direction between **i** and **j**.

e) The resultant force acts in a direction between **i** and $-\mathbf{j}$.

Stop and think answers

1a) West

 b) North

 c) North East

 d) North West

 e) South

 f) South East

2a) $p = 4$ and $q = 2$.

 b) $p = 4$.

 c) $q = 2$

 d) $p < 4$ and $q > 2$

 e) $p < 4$ and $q < 2$

Fig. 6.1
The door.

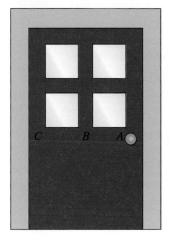

In the section on Statics, we considered the idea of the forces acting on a particle being in equilibrium. When extending the ideas of equilibrium to simple **rigid bodies**, such as a **rod**, the point where the force acts becomes very important. The use of **moments** takes account of the force and where it acts and has to be used in solving equilibrium problems with rigid bodies.

Introducing the idea of a moment of a force

Experiment with a door
Imagine that you are being asked to open a spring loaded door. Consider the door shown in Figure 6.1. You would normally push the door near to the handle at *A*. This point is close to the edge of the door furthest away from the hinges, as shown in Figure 6.2.

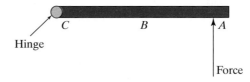

Fig. 6.2
Looking at the door from above

But what if you push at a point in the middle of the door, say at *B*, or at a point even closer to the hinges, at *C*?

When you push the door at a point closer to the hinges it is harder to open. Pushing at *A* is fine, pushing at *B* is harder and when pushing at *C* it can be really quite hard to open the door. The greater the distance between the hinge and the point where you push the easier it is to open the door.

You might like to find a suitable door and try this simple experiment.

We can say that the **turning effect** of the force is greater when it is applied at a point that is further from the hinge, because the door rotates more easily when the same force is applied.

Experiment with a book
Imagine a large book or other heavy object. Place this on the flat of your hand and hold your arm fully outstretched, as shown in Figure 6.3. How does this feel?

Fig. 6.3
Holding the book with your arm outstretched

Fig. 6.4
Holding the book closer to your shoulder.

Fig. 6.5
Holding the book close to your shoulder.

Now bring your hand in to a position closer to your shoulder, roughly half way between the outstretched position and your actual shoulder, as shown in Figure 6.4. How does your arm feel now?

Finally bring your arm in so that your hand is just above your shoulder as in Figure 6.5. How does your arm feel now?

In this example, your arm wants to turn and rotate clockwise about the joint in your shoulder. It is the distance from your shoulder that determines the size of the turning effect.

You will have realised that the turning effect of the weight of the book is greater when your arm is stretched out further and the book is a greater distance from your shoulder.

Definition of a moment of a force

The turning effect considered above can be quantified and is called a **moment**. The moment of the force must be calculated with respect to a particular point and is the product of the **perpendicular distance** between the point and the force and the magnitude of the force.

Magnitude of the moment $= Fd$

where:

$F =$ magnitude of the force

$d =$ perpendicular distance

Here the moment has been calculated with reference to the point O.

Consider again the door example. For example, if applied force is 60 N and it acts at a point 0.5 m from the hinge, as in Figure 6.7, then the moment about the hinge can be calculated.

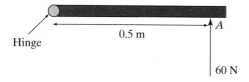

Moment $= 60 \times 0.5 = 30$ Nm (Anticlockwise).

Note that the moment is anticlockwise as the applied force would cause the door to rotate anticlockwise.

Clockwise and anticlockwise moments

As well as having a magnitude or size, a moment of a force will have a direction, either clockwise or anticlockwise according to the rotation it would cause. It is important to consider this when working with moments.

The best way to think about this is to imagine that only the one force acts, and to think about which way the body would rotate when acted on by just this one force if there was a pivot or hinge at the point about which you are taking moments. Figure 6.8 shows two examples.

Fig. 6.6
A force acting at a distance from a point.

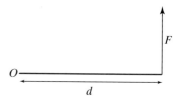

Essential notes

The magnitude of the moment $= Fd$

where:

$F =$ magnitude of the force

$d =$ perpendicular distance.

Fig. 6.7
The door and the force

Fig. 6.8
Anticlockwise and clockwise moments.

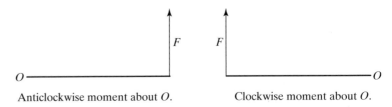

Anticlockwise moment about O.   Clockwise moment about O.

Essential notes

The units used for moments are N m or newton metres. This is because a moment is the product of a force with units N and a distance with units m.

Often positive or negative signs are used when considering the direction of a moment:

- Clockwise moments are negative.

- Anticlockwise moments are positive.

Fig. 6.9
The lever and the force.

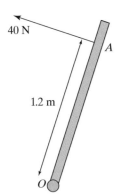

Example

Figure 6.9 shows a lever that is pivoted at the point O. A force of magnitude 40 N acts perpendicular to the lever at the point A, which is 1.2 m from O.

Find the moment of this force.

Answer

$$\text{Moment} = Fd$$
$$= 40 \times 1.2$$
$$= 48 \text{ N m}$$

The moment of this force is 48 N m. When the force is applied as shown in the diagram, it would cause the lever to rotate in an anticlockwise direction about O. As this is in an anticlockwise direction the moment is positive.

Essential notes

The fact that the rod is **uniform** means that the weight must act at the centre of the rod.

Example

A uniform rod has length 4 m and mass 50 kg.

a) Calculate the weight of the rod.

b) Calculate the moment of the weight about the left hand end of the rod.

c) State the moment of the weight about the right hand end of the rod.

Answer

a) Weight $= 50 \times 9.8 = 490$ N

Fig. 6.10
The rod and its weight.

b)

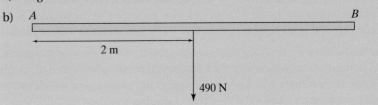

Taking moments about the left hand end of the rod at A, gives:

$$\text{Moment} = Fd$$
$$= 490 \times 2$$
$$= 980 \text{ N m}$$

$$\text{Moment} = -980 \text{ N m}$$

c) Taking moments about B gives 980 N m.

Moments and equilibrium

Equilibrium experiment

Place a book on a flat horizontal surface and apply forces of equal magnitudes and opposite directions as shown in Figure 6.11, so that the forces are applied at the mid-points of opposite sides of the book.

In this case the book remains at rest. The resultant force on the book is zero and the overall moment is zero since the two forces act along the same straight line.

Now try the same thing but without the forces being in line as shown in Figure 6.12.

Here the book starts to rotate. Although the resultant force on the book is still zero, the overall or resultant moment is not zero. For example taking moments about the corner, O, of the book, shown in Figure 6.13, gives:

- The force acting to the right passes through the corner O, so the perpendicular distance is zero and the moment is zero.

- The force acting along the top edge causes an anticlockwise and hence positive moment a perpendicular distance of a, giving a moment of Fa.

$$\text{Total Moment} = Fa - 0a = Fa$$

Any object will only remain in **equilibrium** if the resultant force and the total moment are both zero. Often a good approach is to equate the clockwise and anticlockwise moments.

This result can be used to solve a lot of problems where bodies are in equilibrium. For example if a plank, of known mass, is supported by two concrete blocks, we can find the reaction force exerted on the plank by each of the blocks, R_1 and R_2 in Figure 6.14.

Fig. 6.11
The forces acting on the book are in equilibrium.

Fig. 6.12
The forces acting on the book are not in equilibrium.

Fig. 6.13
Diagram to show taking moments.

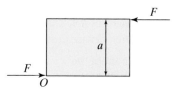

Essential notes

For a rigid body to be in equilibrium the resultant force must be zero **and** the total moment must be zero.

Fig. 6.14
The forces on the plank

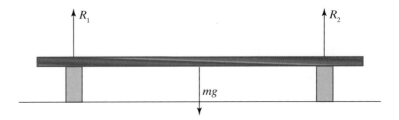

Essential notes

When solving problems with rigid bodies in equilibrium, you can use:

$$\frac{Clockwise}{Moment} = \frac{Anticlockwise}{Moment}$$

Example

A uniform plank of mass 20 kg and length 5 m rests on two concrete blocks. One block is 0.5 m from the left hand end of the plank and the other is 1 m from the right hand end of the plank. Determine the forces exerted on the plank by the blocks.

Answer

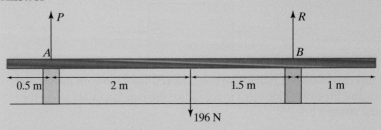

Fig. 6.15
The forces on the plank.

Method notes

Figure 6.15 shows the forces acting on the plank and also the distances between the forces.

Considering the vertical forces gives one equation involving P and Q.

Note that:

The moment of P is zero, as it passes through the point about which moments are taken.

The moment of R is $3.5R$ N m in an anticlockwise direction.

The moment of the weight is $196 \times 2 = 392$ N m in a clockwise direction.

Finally using the equation $P + R = 196$, the value of P can be found.

Vertically:

$$P + R = 196$$

Moments about A:

Clockwise moment = Anticlockwise moment

$$196 \times 2 = 3.5R$$

$$R = \frac{392}{3.5} = 112 = 110 \text{ N (to 2sf)}$$

Finding P:

$$P + R = 196$$
$$P + 112 = 196$$
$$P = 196 - 112 = 84 \text{ N}$$

Example

A uniform rod has mass 20 kg and length 6 m. The rod is held in equilibrium by two vertical ropes that are attached at the points, A and B, which are 1 m from each end of the rod. An object of mass 5 kg is attached by a rope to the left hand end of the rod, as shown in Figure 6.16. Determine the tension in each rope.

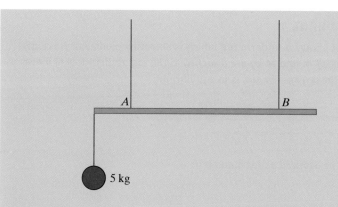

Fig. 6.16
The rod, ropes and mass.

Answer

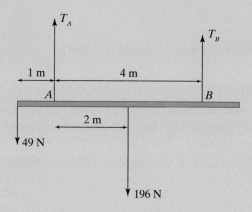

Vertically:

$$T_A + T_B = 49 + 196$$
$$T_A + T_B = 245$$

Moments about A:

Clockwise moment = Anticlockwise moment
$$2 \times 196 = 4T_B + 1 \times 49$$
$$343 = 4T_B$$
$$T_B = \frac{343}{4} = 86 \text{ N (to 2sf)}$$

Returning to the vertical equation:

$$T_A + T_B = 245$$
$$T_A + \frac{343}{4} = 245$$
$$T_A = 245 - \frac{343}{4} = 160 \text{ N (to 2sf)}$$

Fig. 6.17
The forces and distances.

Fig. 6.18
The rod and the particles.

Stop and think 1

Figure 6.14 shows a uniform rod which is smoothly pivoted at its centre, O. Particles, of m and M kg are attached to the rod at distances of a and b m from the centre. The rod is in equilibrium.

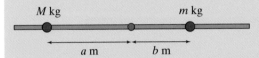

M kg   m kg

a m   b m

What can be deduced about a and b if:

a) $M = m$

b) $M = 2m$

c) $M > m$

d) $4 M = m$

Would it make any difference if the rod was light?

Moments and the perpendicular distance

In all the examples considered so far it has been easy to calculate the moment of a force because the force has always been perpendicular to the specified distance. However, this is not always the case.

Figure 6.19 shows a force, F, which acts at a distance d from the point O.

Fig. 6.19
Finding the perpendicular distance.

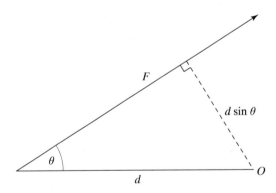

Essential notes

When the distance specified is not perpendicular to the force, the perpendicular distance must be found using trigonometry.

The perpendicular distance for the moment calculation is shown by dashed line in Figure 6.19. This distance can be calculated using trigonometry. In this case:

$$\text{perpendicular distance} = d \sin \theta$$

So the magnitude of the moment is given by:

$$\text{magnitude of the moment} = F d \sin \theta$$

In this case the moment would be negative as it is a clockwise moment. This is because if only the force F acted it would cause a clockwise rotation.

In some cases it is necessary to extend the **line of action of the force**, so that a perpendicular can be drawn from the point about which moments are being taken to the line. Figure 6.20 shows an example where the line of action of the force has been extended with a dashed line and the perpendicular distance has been indicated.

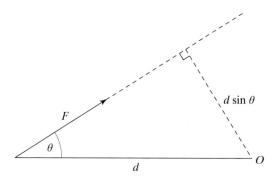

Fig. 6.20
Extending the line of action of a force to find the moment.

Example

A horizontal force of magnitude 120 N is applied to one end of a lever of length 1.5 m whose other end is smoothly hinged at the floor. The force is applied when the lever is at an angle of 60° to the horizontal.

Find the moment of this force about the other end of the lever.

Answer

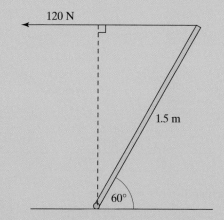

Fig. 6.21
The lever and the force.

Perpendicular Distance $= 1.5 \sin 60°$

Moment $= 120 \times 1.5 \sin 60°$
$= 156$ (to the nearest N m)

Method notes

Figure 6.21 shows the lever and the force. The dashed line is the perpendicular distance which is required here. This is found by using trigonometry in the right angled triangle.

The moment is found by multiplying the force by the perpendicular distance.

Fig. 6.22
The rod and the rope.

Example

Figure 6.22 shows a rod, *AB*, of length 3 m. The rod is held in position by a vertical rope, attached to the rod at *B*. The tension in the rope is 200 N. The rod makes an angle of 30° with the horizontal. Find the moment of the tension about *A*.

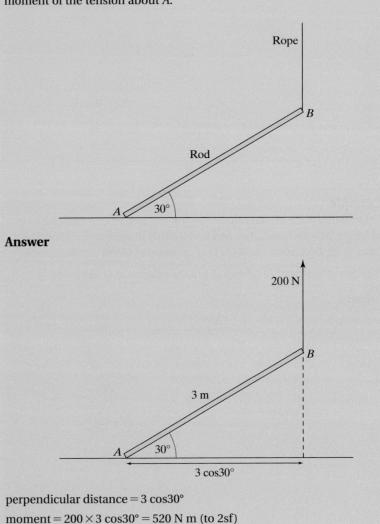

Answer

Fig. 6.23
The forces and the distances.

Method notes

The diagram in Figure 6.23 shows the forces and the distances involved. To find the perpendicular distance the line of action of the force has to be extended, as shown by the dashed line from *B*.

Using trigonometry the perpendicular distance from *A* to this line can be calculated.

Finally use moment = force × perpendicular distance.

perpendicular distance = 3 cos30°
moment = 200 × 3 cos30° = 520 N m (to 2sf)

Non-uniform rods

For a non-uniform rod the weight is assumed to act at a point known as the **centre of mass** rather than at the centre of the rod. If the position of the centre of mass is known, then the weight force can be assumed to act at this point and moments can be used in a similar way to the examples above. If the forces acting on the non-uniform rod are known, then these can be used to find the position of the centre of mass.

Example

A non-uniform rod, *AB*, has length 6 m and weight 75 N. It is suspended in equilibrium in a horizontal position by vertical ropes attached to each end of the rod. The tension in the rope attached at *A* is 50 N.

a) State the tension in the rope attached at *B*.

b) Find the distance of the centre of mass of the rod from *A*.

Answer

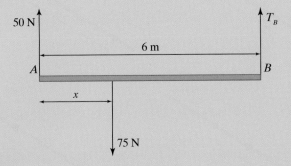

Fig. 6.24
The forces on the non-uniform rod

a) Vertically:

$$50 + T_B = 75$$
$$T_B = 25 \text{ N}$$

b) Taking moments about *A*:

$$75x = 25 \times 6$$
$$x = \frac{150}{75} = 2 \text{ m}$$

The centre of mass is 2 m from *A*.

Method notes

The diagram in Figure 6.24 shows the forces and where they act. The weight has been shown acting at a distance of *x* m from *A*.

The unknown tension is found in part (a) by considering the fact that vertical forces are in equilibrium.

In part (b) moments are taken about the point *A*. The moment of the tension at *B* (25×6) is anti-clockwise and the moment of the weight ($75x$) is clockwise. These moments can be used to form an equation to find *x*.

Fig. 6.25
Pivoted rods and particles.

Stop and think 2

Each of the diagrams shown in Figure 6.25 shows a uniform rod which is smoothly pivoted at its centre. Particles have been attached to the rod as shown in each diagram and the rod is released from rest in the positions shown. In each case determine if the rod will remain at rest or rotate.

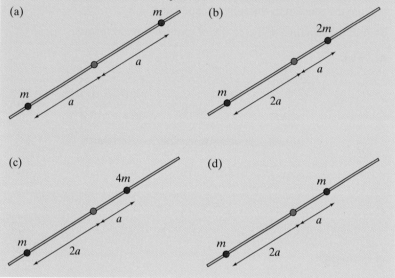

Stop and think answers

1a) $a = b$

b) $b = 2a$

$\quad b > a$

$\quad a = 4b$

It would make no difference if the rod was light.

2a) Remains at rest

b) Remains at rest

c) Starts to rotate clockwise

d) Starts to rotate anticlockwise

Questions

1. Two particles, A and B, are moving towards each other on a straight horizontal line. The speed of A is 8 m s^{-1} and the speed of B is 4 m s^{-1}. The mass of A is 7 kg and the mass of B is m kg. The particles collide. Immediately after the collision, the direction of motion for both particles has changed. Particle A moves with a speed of 5 m s^{-1} and particle B has speed 3 m s^{-1}.

   a) Find the magnitude of the impulse on A during the collision. (2)

   b) Find m. (3)

2. A student completes a 100 m sprint in a total of 13.5 seconds. A simple model assumes that the student accelerates uniformly for 2 seconds to reach a speed of V m s^{-1} and then moves at a constant speed of V m s^{-1} to complete the race.

   a) Sketch a velocity-time graph. (3)

   b) Find V. (3)

3. A particle is fired vertically upwards with at a speed of 21 m s^{-1} form a point at a height of 17.5 m above ground level

   a) Find the maximum height of the particle above ground level. (3)

   b) Find total time from when the ball is fired to when it hits the ground for the first time. (5)

4. A uniform rod, AB has length 4 m and mass 20 kg. Two ropes are attached to the rod. One is attached at A and the other at the point C, where $AC = 2.5$ m, as shown in Figure 1. The rod remains in equilibrium in a horizontal position.

Fig. 1

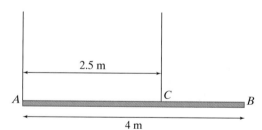

a) Show that the tension in the rope attached at C is 16g. (3)

b) Find, in terms of g, the tension in the rope attached at A. (2)

c) A particle of mass m is attached at B. The tension in the rope at C is now 8 times greater than the tension in the rope at A. Find m. (7)

5. A box of mass 30 kg, remains at rest on a rough horizontal surface, when a force of magnitude 80 N is applied as shown in the diagram in Figure 2.

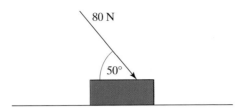

Fig. 2

Model the box as a particle.

a) Find the magnitude of the normal reaction force acting on the box. (3)

b) If the box is on the point of sliding, find the coefficient of friction between the box and the surface. (4)

6. A van, of mass 2400 kg, is used to tow a trailer, of mass 1600 kg. They accelerate uniformly along a straight horizontal road, from rest and reach a speed of 20 m s^{-1} when they have travelled 80 m.

a) Find the acceleration of the van and trailer. (2)

Resistance forces act on both the van and the trailer. The magnitude of the resistance force on the van is twice the magnitude of the resistance force on the trailer. A forward driving force of magnitude 11050 N acts on the van.

b) Find the magnitude of the resistance force acting on the van. (4)

c) Find the magnitude of the horizontal force that the van exerts on the trailer. (3)

Later the driver of the van applies the brakes. The van and the trailer then start to decelerate at 0.5 m s^{-2}. Assume that the resistance forces on both vehicles remain the same.

d) Find the magnitude of the horizontal force that the trailer exerts on the van and state the direction in which it acts. (3)

7. A particle, of mass 5 kg, is placed on a plane inclined at an angle of 20° to the horizontal. A horizontal force of magnitude X N acts on the particle, as shown in Figure 3. The particle accelerates up the plane at 1.2 m s^{-2}.

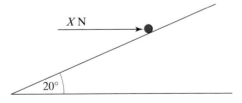

Fig. 3

a) Assume that the plane is smooth and find X. (4)

The slope is in fact rough and that the coefficient of friction between the slope and the particle is 0.4.

b) Find X. (7)

8. In this question, **i** and **j** are horizontal unit vectors directed east and north respectively. All the position vectors are given with respect to a fixed origin O.

   A ship moves from a point with constant velocity **v** km h^{-1}. At noon it has position vector 230**i** + 60**j** km and at 8.00pm it has position vector 290**i** − 20**j** km.

   a) Find **v**. (3)

   A boat leaves the point with position vector 360**i** −110**j** km at 8.00pm to intercept the ship. It moves with a constant velocity of −6.5**i** + 8**j** km h^{-1}.

   b) Find the distance between the boat and the ship at 10.00pm. (6)

   c) Show that the boat intercepts the ship and find the time when this happens. (6)

Answers

1. (a) $I = 7 \times (-5) - 7 \times 8$ (2)
   $$= -35 - 56$$
   $$= -91 \text{ N s}$$
   $$\therefore \text{Magnitude} = 91 \text{ N s}$$

   (b) $7 \times 8 + m \times (-4) = 7 \times (-5) + m \times 3$ (3)
   $$56 - 4m = 3m - 35$$
   $$7m = 91$$
   $$m = \frac{91}{7} = 13 \text{ kg}$$

2. (a)   (3)

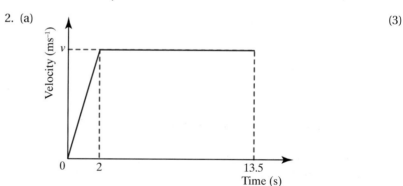

   (b) $\frac{1}{2} \times 2 \times V + 11.5 \times V = 100$

   $$12.5V = 100$$ (3)

   $$v = \frac{100}{12.5} = 8 \text{ m s}^{-1}$$

3. (a) $v^2 = u^2 + 2as$ (3)
   $$0^2 = 21^2 + 2 \times (-9.8)s$$
   $$s = \frac{21^2}{2 \times 9.8} = 22.5 \text{ m}$$
   $$\text{Height} = 22.5 + 17.5 = 40 \text{ m}$$

(b)
$$s = ut + \frac{1}{2}at^2 \qquad (5)$$

$$-17.5 = 21t + \frac{1}{2} \times (-9.8)t^2$$

$$4.9t^2 - 21t - 17.5 = 0$$

$$t = \frac{21 \pm \sqrt{21^2 - 4 \times 4.9 \times (-17.5)}}{2 \times 4.9}$$

$$= 5 \text{ or } -0.714\ldots$$

Total time = 5 seconds

OR $\quad v^2 = u^2 + 2as$

$$v^2 = 21^2 + 2 \times (-9.8) \times (-17.5)$$

$$v = \sqrt{784} = \pm 28$$

$$v = u + at$$

$$-28 = 21 + (-9.8)t$$

$$t = \frac{21 + 28}{-9.8} = 5 \text{ seconds}$$

OR

Upward motion:

$$v = u + at$$

$$0 = 21 - 9.8t$$

$$t = \frac{21}{9.8} = 2.143$$

Downward motion:

$$s = ut + \frac{1}{2}at^2$$

$$40 = \frac{1}{2} \times 9.8t^2$$

$$t = \sqrt{\frac{40}{4.9}} = 2.857$$

Total time = 2.143 + 2.857 = 5 seconds

4. (a) Moments about A: $\qquad (3)$

$$T_C \times 2.5 = 20g \times 2$$

$$T_C = \frac{40g}{2.5} = 16g$$

(b) $20g = 16g + T_A \qquad (2)$

$$T_A = 4g$$

(c) Moments about A: $\qquad (7)$

$$T_C \times 2.5 = 20g \times 2 + mg \times 4$$

$$T_C = \frac{40g + 4mg}{2.5}$$

$$T_A = \frac{T_C}{8}$$

$$T_A + T_C = 20g + mg$$

$$\frac{T_C}{8} + T_C = 20g + mg$$

$$\frac{9}{8}T_C = 20g + mg$$

$$\frac{9(40g + 4mg)}{8 \times 2.5} = 20g + mg$$

$$360g + 36mg = 400g + 20mg$$

$$16m = 40$$

$$m = \frac{40}{16} = 2.5 \text{ kg}$$

5. (a) $R = 80 \sin 50° + 30 \times 9.8$ (3)

$$= 355.2 \ldots = 360 \text{ N (to 2sf)}$$

(b) $F = 80 \cos 50°$ (4)

$$F = \mu R$$

$$80 \cos 50° = \mu(80 \sin 50° + 294)$$

$$\mu = \frac{80 \cos 50°}{80 \sin 50° + 294} = 0.14 \text{ (to 2 sf)}$$

6. (a) $20^2 = 0^2 + 2 \times a \times 80$ (2)

$$a = \frac{20^2}{2 \times 80} = 2.5 \text{ m s}^{-2}$$

(b) $11050 - 3R = (2400 + 1600) \times 2.5$ (4)

$$R = \frac{11050 - 10000}{3} = 350 \text{ N}$$

(c) $T - 350 = 1600 \times 2.5$ (3)

$$T = 4350 \text{ N}$$

OR

$$11050 - 2 \times 350 - T = 2400 \times 2.5$$

$$T = 11050 - 700 - 6000 = 4350 \text{ N}$$

$$T - 350 = 1600 \times (-0.5)$$ (3)

$$T = -800 + 350 = -450$$

Magnitude $= 450$ N in direction of motion

7. (a) $X \cos 20° - 5 \times 9.8 \sin 20° = 5 \times 1.2$ (4)

$$X = \frac{5 \times 1.2 + 5 \times 9.8 \sin 20°}{\cos 20°} = 24 \text{ N (to 2 sf)}$$

(b) Perpendicular to the slope: (7)

$$R = 49 \cos 20° - X \sin 20°$$

Parallel to the slope:

$X \cos 20° - F - 5 \times 9.8 \sin 20° = 5 \times 1.2$

$X \cos 20° - 0.4(49 \cos 20° + X \sin 20°) - 49 \sin 20° = 6$

$X(\cos 20° - 0.4 \sin 20°) = 6 + 49 \sin 20° + 19.6 \cos 20°$

$X = \dfrac{6 + 49 \sin 20° + 19.6 \cos 20°}{\cos 20° - 0.4 \sin 20°} = 51 \text{ N (to 2sf)}$

8. (a) $290\mathbf{i} - 20\mathbf{j} = 230\mathbf{i} + 60\mathbf{j} + 8\mathbf{v}$ (3)

$\mathbf{v} = \dfrac{60\mathbf{i} - 80\mathbf{j}}{8} = 7.5\mathbf{i} - 10\mathbf{j}$

(b) $\mathbf{r}_s = 290\mathbf{i} - 20\mathbf{j} + 2(7.5\mathbf{i} - 10\mathbf{j}) = 305\mathbf{i} - 40\mathbf{j}$ (6)

$\mathbf{r}_B = 360\mathbf{i} - 110\mathbf{j} + 2(-6.5\mathbf{i} + 8\mathbf{j}) = 347\mathbf{i} - 94\mathbf{j}$

$\mathbf{r}_S - \mathbf{r}_B = -42\mathbf{i} + 54\mathbf{j}$

Distance $= \sqrt{42^2 + 54^2} = 68 \text{ km (to the nearest km)}$

(c) $\mathbf{r}_s = 290\mathbf{i} - 20\mathbf{j} + (7.5\mathbf{i} - 10\mathbf{j})t = (290 + 7.5t)\mathbf{i} + (-20 - 10t)\mathbf{j}$ (6)

$\mathbf{r}_B = 360\mathbf{i} - 110\mathbf{j} + (-6.5\mathbf{i} + 8\mathbf{j})t = (360 - 6.5t)\mathbf{i} + (-110 + 8t)\mathbf{j}$

$290 + 7.5t = 360 - 6.5t$ and $-20 - 10t = -110 + 8t$

$14t = 70$ $90 = 18t$

$t = 5$ $t = 5$

$\therefore$ Interception takes place 1.00 am

Miscellaneous symbols

$=$	is equal to
$\neq$	is not equal to
$\equiv$	is identical to or is congruent to
$\approx$	is approximately equal to
$<$	is less than
$\leq$	is less than or equal to, is not greater than
$>$	is greater than
$\geq$	is greater than or equal to, is not less than
∞	infinity

Operations

$a + b$	a plus b
$a - b$	a minus b
$a \times b$, ab, $a.b$	a multiplied by b
$a \div b, \dfrac{a}{b}, a/b$	a divided by b
$\sqrt{a}$	the positive square root of a

Vectors

$\mathbf{i}, \mathbf{j}, \mathbf{k}$	unit vectors in the directions of the cartesian coordinate axes

Mechanics

Constant Acceleration

u	initial velocity
v	final velocity
t	time
a	acceleration
s	displacement
g	acceleration due to gravity

Friction Law

F	friction
μ	coefficient of Friction
R	normal reaction

Newton's Second Law

F	resultant Force
m	mass
a	acceleration

Impulse

I	impulse
m	mass
t	time
u	initial velocity
v	final velocity

Conservation of Momentum

m_A	mass of A
m_B	mass of B
u_A	initial velocity of A
u_B	initial velocity of B
v_A	final velocity of A
v_B	final velocity of B

Displacement

v	velocity
t	time

Position Vector for Constant Velocity

$\mathbf{R}$	position vector
$\mathbf{v}$	velocity
t	time
$\mathbf{r}_0$	initial position

Moment

F	magnitude of the force
d	perpendicular distance

The formulae listed below have been used throughout the book and many will be required in the examination. These formulae are not listed in the formula booklet that will be available in the examination, so it is important that you learn these and can state them quickly and easily in your examination.

Constant Acceleration Equations

u = initial velocity

v = final velocity

t = time

a = acceleration

s = displacement

$$v = u + at$$

$$s = \frac{1}{2}(u + v)t$$

$$s = ut + \frac{1}{2}at^2$$

$$s = vt - \frac{1}{2}at^2$$

$$v^2 = u^2 + 2as$$

Quadratic Equation Formula

$$ax^2 + bx + c = 0$$

$$x = \frac{-b \pm \sqrt{b^2 - 4ac}}{2a}$$

Weight

Weight = mg

Friction Law

F = friction

μ = coefficient of Friction

R = normal reaction

$$F \le \mu R$$

Newton's Second Law

F = resultant Force

m = mass

a = acceleration

$$F = ma$$

Momentum

Momentum = Mass $\times$ Velocity

Impulse

I = impulse

m = mass

t = time

u = initial velocity

v = final velocity

$I = mv - mu$

$I = Ft$

Conservation of Momentum

m_A = mass of A

m_B = mass of B

u_A = initial velocity of A

u_B = initial velocity of B

v_A = final velocity of A

v_B = final velocity of B

$m_A u_A + m_B u_B = m_A v_A + m_B v_B$

Total Momentum Before = Total Momentum After

Sine and Cosine Rules

$$\frac{a}{\sin A} = \frac{b}{\sin B} = \frac{c}{\sin C}$$

$$a^2 = b^2 + c^2 - 2bc \cos A$$

Lami's Theorem

$$\frac{P}{\sin \alpha} = \frac{Q}{\sin \beta} = \frac{R}{\sin \gamma}$$

The magnitude of the vector $a\mathbf{i} + b\mathbf{j}$

$$\sqrt{a^2 + b^2}$$

Displacement

v = velocity

t = time

Displacement = vt

Position Vector for Constant Velocity

$\mathbf{R}$ = position vector

$\mathbf{v}$ = velocity

$t = $ time

$\mathbf{r}_0 = $ initial position

$\mathbf{r} = \mathbf{v}\,t + \mathbf{r}_0$

Unit vector in the direction of $a\mathbf{i} + b\mathbf{j}$

$$\frac{a\mathbf{i} + b\mathbf{j}}{\sqrt{a^2 + b^2}}$$

Constant Acceleration equation

$\mathbf{u} = $ initial velocity

$\mathbf{v} = $ final velocity

$t = $ time

$\mathbf{a} = $ acceleration

$\mathbf{v} = \mathbf{u} + \mathbf{a}t$

Newton's Second Law for Vectors

$\mathbf{F} = $ resultant Force

$m = $ mass

$\mathbf{a} = $ acceleration

$\mathbf{F} = m\mathbf{a}$

Moment

$F = $ magnitude of the force

$d = $ perpendicular distance

The magnitude of the moment $= Fd$

Index